ALSO BY BOB BERMAN

Cosmic Adventure

Secrets of the Night Sky

Strange
Universe

Strange Universe

THE WEIRD AND WILD SCIENCE OF
EVERYDAY LIFE—ON EARTH AND BEYOND

BOB BERMAN

TIMES BOOKS | Henry Holt and Company
New York

Times Books
Henry Holt and Company, LLC
Publishers since 1866
115 West 18th Street
New York, New York 10011

Henry Holt ® is a registered trademark of
Henry Holt and Company, LLC

Library of Congress Cataloging-in-Publication Data

Berman, Bob.

Strange Universe / Bob Berman.—1st ed.

p.   cm.

Includes index.

ISBN 0-8050-7328-0

1. Astronomy—Popular works. I. Title.

QB44.3.B37 2003

520—dc22   2003060687

Henry Holt books are available for special
promotions and premiums.
For details contact:
Director, Special Markets.

First Edition, 2003

Designed by Paula Russell Szafranski

Printed in the United States of America

1   3   5   7   9   10   8   6   4   2

To the women of my family
(in order of increasing distance from my word processor):
Marcy, Anjali, Alyssa, Jane, Rita, Paula, Lara, Dorianne, and Jessica.

Contents

Strange
Universe

Introduction

The basic realities of everyday existence—light and time and gravity—are astounding if not downright mysterious. Commonplace facets of the physical universe that we experience without thought become magical, marvelous, and even preposterous upon reflection and examination.

Can something as ordinary as a shadow engage us? Consider: Why are tree shadows blue when cast upon snow? What time of every day does Earth's own shadow hover vividly against the sky, ignored by virtually everyone?

Nature's curiosities are everywhere. Doesn't it seem odd (as it did to the ancient Greeks) that the full moon looks flat as a dish, as if painted onto the heavens? It baffled moonwatchers through the ensuing centuries—until just recently, when the flat-moon riddle was solved.

The briefest inquiry into even ordinary substances can rivet the mind. Case in point: water. It's the most common compound in the cosmos. And yet in liquid form it exists in only one other place in the known universe. Given the tiny size of its molecule, water ought to be a gas at room temperature, yet it isn't, thanks to a single reality-altering quirk. And it absorbs red light but leaves blue and green alone, dyeing oceans turquoise. It slows light to 75 percent of its customary speed, making fish appear in phantom positions. Such unexpected consequences, paradoxes, and improbable truths await the *Eureka!* of our notice.

The strange objects, events, processes, and phenomena described here have been selected for their everyday familiarity. My goal is not to dumb down complexities or explain well-known oddities to the scientifically challenged, but rather to offer readers, teachers, and other professionals a harvest of fresh material to share with friends and students. Spanning a wide and random range, the subjects have been organized into two sections. The first embraces earthly peculiarities of our lives and our world: surprising historical lore, statistical peculiarities, unexamined events like Groundhog Day, the vagaries of gravity, the nature of light. The second looks at anomalies and mysteries beyond our planet, from previously unpublished blunders and antics of the U.S. and Russian space programs to the smells of space to the Big Bang itself.

A comprehensive, thorough exposition of the universe's curiosities would require many thick volumes. The choice of topics is based on recent scientific revelations and my own sense of what is likely to fascinate readers. After three decades of teaching college science courses, and as a longtime columnist for *Discover* magazine and an editor and columnist for *Astronomy* magazine and the *Old Farmers Almanac,* I've learned that the universe is as full of curveballs as it is of electrons. If I've succeeded, the book follows that model, illuminating both minuscule and monumental realities as ordinary as a snowball and as strange as an apple seed that outweighs an aircraft carrier.

I've sought to convey what I've learned of things hilarious and horrendous, the goofs and the glories, the magnificent harmony and the significant chaos—the puzzles and patterns comprising our daily lives and our whirling world.

Join me in a rousing journey full of discoveries that I hope will "make your day" every day of your life—as they have mine.

Part I

What's Going On Here?

Chapter 1

Look Out Below!

Only a glitch in the laws of chance can explain why all the major meteor showers arrive in a single twenty-week span. During the annual shooting-star fireworks that peak around August 12, October 21, November 18, December 13, and January 3, up to a hundred bits of cometary debris visibly slice through the air just above us every hour. But in early winter the displays end. For seven straight months, little more than sporadic pellets pepper the heavens. So if your personal paranoia extends all the way into space and your HMO doesn't cover you for meteor assault, you'd assume it's safe to relax until midsummer.

No such luck. History tells us that whenever a person, home, or property has suffered a close encounter of the clobbering kind it's happened during non-shower periods. Which makes sense when you think about it. Except for December 13's Geminids, composed of sturdier asteroid fragments, every major shower is made of skimpy, apple-seed-size rubble from comets, lightweight material that burns to oblivion during its collision with our atmosphere. Only a dusty trail lingers like a Cheshire Cat's smile after the intruder has gone to meteor heaven. Sporadics, however, can be made of anything and come from anywhere. They can be rocky asteroid fragments or even bits of the Moon or Mars. Many contain substantial amounts of iron or nickel, and depending on their speed and initial size, some of this hardy stuff makes it all the way to the ground.

Slowed and cooled by the thick lower atmosphere, meteors no longer incandesce as they approach the surface and are barely warm when they land. On August 31, 1991, one smashed into a lawn next to two boys in Noblesville, Indiana, who immediately picked it up and later described it as slightly warm to the touch. In fact, meteors are not hot enough to glow once they've dipped below fifty miles in altitude, having decelerated dramatically from their original fierce speed of twelve to forty-five miles per second down to anywhere from a twentieth to half a mile per second. But that's still fast enough to cause trouble.

Discover magazine's "Risk" issue (May 1996) presented statistical evidence that we're each six times more likely to die from a meteor impact than in an airliner crash. The reason: The post-impact effects of a meteor a few miles wide can wipe out most or all of the human race, while relatively few of us will ever be crash victims.

Virtually all meteors that land or airburst produce some damage, even if they don't remotely compare to monsters that come in on longer time scales and prompt serious philosophical questions about the general stability of the solar system. The most famous such humongous impact destroyed half the life-forms of Earth 65 million years ago, at the end of the Cretaceous period and the dawn of the Tertiary era—known familiarly to geologists and meteor buffs as the K-T boundary. The resulting undersea crater, some 120 miles across and located just off Mexico's Yucatán peninsula, testifies to the amount of material blown skyward. It blackened the air for years, erased the dinosaurs as thoroughly as a studio's special-effects department, and allowed us mammals to begin our spectacular evolutionary ascent toward rats and sitcoms.

A lesser known but far greater disaster happened 185 million years before that, at the end of the Permian period. In 2002, a team of geochemists led by Tokyo University's Kunio Kaiho uncovered convincing evidence that this greatest mass extinction of all time, which dwarfed the K-T event, destroyed 95 percent of all earthly species. Corroborating recent studies that examined trapped gases from sediments in various parts of the world, Kaiho discovered that the impact also converted solid sulfur into so much sulfur-rich gas that up to 40 percent of Earth's atmosphere was consumed. The resulting frenzy of global acid production pickled the oceans and raised the acidity of the sea's surface to that of lemon juice! Earth was nearly dealt a knockout blow. Plants and microscopic animals that had survived for a million centuries suddenly disappeared forever.

Lesser violence has been far more frequent if not nearly as damaging, chronicled by craters like the Barringer colossus in the desert west of Winslow, Arizona, a hole almost a mile wide created 50,000 years ago.

Nor has the violence been limited to ancient history. Some nonagenarians were alive on June 30, 1908, when more than 500 square miles of less-than-fashionable real estate in Siberia's Tunguska Basin was flattened by a large air-bursting meteor. A man fifty miles away was knocked from his chair by the force of the blast, but no one else in that virtually uninhabited region was injured. Even the damage went unrecorded until explorers, reaching the area a decade later, gaped at the spectacle of countless downed conifers pointing radially outward from ground zero.

The real troublemakers (say, one to five miles across) rarely clobber us, simply because there's so much room around our planet—which is why it's called space. They are becoming disconcerting, however: Increasing ability to detect them is showing how frequently we escape. In 2002, a Tunguska-class asteroid missed us by less than half the distance to the Moon.

Every 1,000 years or so when a big one arrives, the only good news is that most of our planet's area is oceanic or uninhabited; each time, the odds are in our favor. In between, we experience the *bam-bam-bam* of smaller strikes, which forever fall among us as if hurled by a relentless cosmic baseball-throwing machine. A building in North America is struck every fifteen months, on average.

Sounds ominous. But often a meteor brings temporary celebrity and even wealth. And—so far—never grievous injury, unless one includes the Franciscan monk in Milan who died mysteriously in the seventeenth century after being struck in the leg by a plum-sized stone "falling from the sky" which severed an artery. While possibly true, the account is not confirmable; nor are newspaper reports of a calf struck in Ohio in 1860 and a dog killed in Egypt in 1911.

In our era, the surest instance of injury to a human or animal is reported in a 1954 *Life* magazine photo-story about a woman in Sylacauga, Alabama. Mrs. E. Hewlett Hodges had been lounging on her couch when a meteor smashed through the roof, ricocheted off her radio, and struck her upper leg. The resultant bruise remains the only confirmed meteorite injury.

Note: Are "meteorite" and "meteor" synonyms? Not at all. A sky streak is always a meteor, and a stone found on the ground is always a meteorite, but the names are interchangeable in only one circumstance: at the moment of collision. Because a meteor becomes a meteorite when it strikes our planet, you can use either term when describing or witnessing an impact. Beyond Earth, the silent, dark, tail-less flying stones have yet another name—meteoroids, which would be the correct term for anything that punches a hole in the Space Station or the shuttle.

Eighteen-year-old Michelle Knapp of Peekskill, New York, didn't believe it herself when she heard a loud crunch outside her house on October 9, 1992. Finding the back of her Chevy grotesquely destroyed, she attributed the damage to some ultrapowerful vandal. She couldn't know that people up and down the East Coast had just seen and videotaped fiery northward-whizzing meteor fragments, and that her car was the lucky recipient of the only piece known to have landed. It didn't seem lucky at first: The damage lay decisively outside the manufacturer's warranty. Even when investigators found the twenty-six-pound meteorite under the car's twisted trunk, the event seemed a bizarre fable in which Destiny had swapped her car for an ugly rock. The happy ending came when she accepted a collector's $69,000 offer for both the meteorite and the totaled ten-year-old Malibu.

Nor did Bob and Wanda Donohue of Wethersfield, Connecticut, feel particularly fortunate on the evening of November 30, 1982. They'd been watching TV (*M*A*S*H*) when the next room suddenly exploded in chaos. Rushing through the door, they found smoke and dust filling the air, furniture knocked over, and a hole in the ceiling. When emergency personnel arrived, a fireman found a six-pound rock under the table, where it had settled after a few floor-to-ceiling bounces chronicled by scuff marks in the carpet. But insurance covered the damage, and the Donohues generously donated the valuable specimen to a New Haven museum.

They weren't quite as amazed as residents of any other town might have been. Eleven years earlier, in April 1971, the previous meteor known to crash into the roof of a house in the United States had struck barely more than a mile from the Donohues. The freak coincidence may be explainable by Wethersfield's proximity to Hartford, with its high concentration of insurance actuaries—experts who would stake their reputation that the same town cannot be hit consecutively by shooting stars.

And it never stops. Several stray meteors will pass within a few dozen miles of you during the next sixty minutes—with twice as many in the A.M. as in the P.M. hours, because you're then on the leading side of Earth as it speeds through space. You've got as good a chance as anyone else to have an asteroid fragment plop onto your kitchen table, bearing the gifts of damage, wealth, and proof that we do indeed live in a strange universe.

Chapter 2

Physics in the Morning

You get up, shower, gulp some breakfast, hop in a car or train, and head off to work.

Science? Not much. So let's replay the routine as if we were as fascinated by the marvels of existence as Archimedes or Newton. First, on awakening, glance at the window. If the air conditioning has been on, mist steaming up the *outside* of the glass means that the dewpoint (the temperature at which dew or fog forms) is higher than the room's temperature, which happens only in extremely humid weather. Moisture on exterior glass is routine in southeastern states, occasional in the Northeast and Midwest, uncommon elsewhere. Bad news. Dress for a sticky day. Even if cloudless, the sky will be milky, not blue.

Does the curtain pull itself inward toward your leg when you turn on the shower? Jets of water from the showerhead will produce this effect: It's Bernoulli's principle. A rapid flow of liquid or gas reduces air pressure in its vicinity. The stream pulls adjacent air along with it, creating a partial vacuum that sucks in the curtain. The same principle makes tornadoes lift roofs and lets airplanes fly.

Shower over, you're warmer if you stay in the stall than if you step out. It's not because the room is cooler but because it's drier. Venturing into lower humidity accelerates moisture evaporation from your skin, and evaporation is always a cooling process because the transition from a liquid to a gaseous phase requires energy. So the leftover water on your skin pulls heat from your body in order to vaporize, chilling you. The refrigerator in your kitchen operates on the same principle.

If it's raining or snowing on your drive to work, watch how all the precipitation seems to come from a single spot directly ahead of you and higher than your eye-level. The faster your speed, the lower the spot from which the shower emanates. This effect is called aberration—a shift in the apparent position of an object due in part to the observer's motion—and it also applies to starlight reaching Earth. All the universe's objects seem displaced a bit from their true position because of Earth's orbital motion. The shift in star position is as much as 20 arcseconds, about the size a grapefruit appears when it's a mile away. (More about aberration in chapter 30.) The radiating point of the rain or snow rises and falls when you brake or accelerate. With a little practice you can accurately determine your speed by the height of this almost hypnotic emanating spot, no speedometer needed.

If the sky is clear, appraise its color. Most people shrug it off; it's blue, so what? But the different shades tell a tale. That the sky gets lighter near the horizon is such an ordinary observation that we scarcely pay attention to it. Yet it gives you an instant environmental report of dust, humidity, and pollution. Looking to the horizon means sighting through thirteen times more air than when gazing straight up. So whatever junk is mixed in gets amplified. Contaminants reflect all the Sun's wavelengths equally, exposing their presence with a whitening effect. Seen from dry, clean sites, the horizon sky barely pales. So the degree of brightening near your skyline is an excellent measure of the air quality of your location; white in the sky denotes filth (or humidity or both).

An emergency vehicle coming at you, siren wailing, will display the wonderful Doppler shift. As it passes you, a double phenomenon suddenly kicks in: The siren's pitch lowers and the spacing between warbles is longer, as if the siren's batteries were wearing out. The sound waves from a fast-approaching police car may strike your ear at 840 mph, while the retreating vehicle's sound arrives at only 620 mph—proof that sound, unlike light, is not a constant. Listen for the same dramatic effect from planes flying overhead.

Stopped in traffic? If you're wearing sunglasses, check out any white clouds near the Sun. Their fringes often display iridescence—vivid pinks, purples, or aquas. The Sun emits no white light; it radiates all the colors of the spectrum. Many surfaces, such as clouds and beaches, reflect all these hues equally, so they reach our eyes and brain together: The scrambled mixture is perceived as white. But since cloud borders often contain rapidly evaporating water droplets of varying diameters, light passing through them travels disparate distances, letting the crest of one light wave coincide with the trough of another, canceling out that light entirely. Removing a major color takes away part of the mix that makes a cloud white. The remaining waves concoct new colors not part of the normal spectrum. Colors produced by interference (like pink and turquoise) are not present in the spectrum of the Sun and the stars; we see these colors nowhere else in the universe.

Similar interferences create those oily colorful swirls of psychedelic design in roadside puddles. Light here is acting like a wave and not a particle, since particles can't interfere in this way. In debates over whether light is a wave or a particle puddle gazers have an advantage.

At the next stoplight, another revelation: Notice how metals around you (like the bumpers of all those trucks) gleam. What's happening when things gleam? Interesting process: Metals have outer electrons capable of absorbing and then re-emitting photons of light, while their inner electrons are frozen in place, with too little flexibility to vibrate and emit photons. (A photon is light in its particle guise.) Result: Sunlight hitting metals does not get past those outer electrons but is reflected from them, making metals neither transparent nor dull but something else: gleamy.

Next stop, take another look at the sky to see if you can spot the Moon. During one week each month you'll easily see it in the morning sky. It will never be full and will always be lit up on the left side. This is the waning Moon, usually a half-moon or nearly so, the phase that lies in front of us as Earth and Moon together orbit the Sun. Look its way and you're in the pilot seat, facing ahead as our planet hurtles through space. When the morning Moon is in front of you, you're looking forward in our orbit and going 66,000 miles per hour over the speed limit.

As you drive, the inside of your car will get warmer, for two reasons. First, humans each radiate an average of 96.8 watts of heat, about the same as a 100-watt bulb. You and a few passengers in an enclosed space will warm it up quickly. The second reason involves sunlight interacting with your car's windows, and reveals the glass's atomic structure. Visible light vibrates about 100 trillion times a second, very nearly matching the natural oscillations of the electrons in window glass. So they readily become stimulated by visual light; each time they're struck by a photon they produce an identical one, which flies on to the next atom and then the next, in an unbroken chain, until a final photon exits the pane and enters the car. This is what makes glass transparent. A photon striking a window is continually reproduced until a perfect copy emerges. The light hitting your eyes is not the same that first struck the glass, but a clone.

Once inside the car, the light is partially absorbed by the seat, your skin, clothing, everything around you, all of which re-emits some of that energy in the infrared (heat) part of the electromagnetic spectrum. But this heat can't readily escape back out: Infrared's long wavelengths cannot make electrons vibrate. Instead they cause entire atoms to oscillate, which creates heat but no new photons. This is why glass is opaque to infrared. It also explains how a greenhouse works: Visible light goes in but the heat doesn't get back out; it builds up inside, same as in your car—and, for that matter, on the broiling planet Venus.

On a hot day, and especially if you're late for work, you might see a mirage on the road, a pool of water that isn't there. In the very hot air just above the asphalt, the speed of light is faster than that in the cooler, denser air a few inches higher up. When light changes speed (it is constant only in a vacuum) it bends, or refracts, creating imaginary reflective zones that look like puddles. Refraction makes a spoon in a half-glass of water look bent, too. It causes fish in pools or tanks to appear in different places from where they really are. It makes that glass cola bottle in your hand seem to hold twice as much as it really does.

You turn on the car radio, and the FM stations fade in and out as you pass close to hills or buildings. But the AM signals are much steadier. This is because of the bending of electromagnetic radiation around obstacles, a phenomenon called diffraction. Longer wavelengths diffract more readily. AM stations broadcast waves far longer than those of the megahertz FM band, so they bend around obstacles much more easily and manage to reach you. Very long waves (like waves on the open ocean) will scarcely be stopped by a small barrier. Notice how waves encountering a little rocky lighthouse island fill in again behind the island and continue on their way. If sea waves were faster and closer together, there would be more of a "shadow zone" behind small islands in which the ocean was permanently calm. Diffraction everywhere!

As soon as you're out of traffic and moving quickly, roll down the window and see how all your important loose papers get sucked out. By now you know why. Bernoulli's principle again. Your car's motion has produced a stream of rapidly moving air just outside the window. Rapid motion means low pressure, which creates enough of a vacuum to pull out the documents. Look through the rear window and watch them blow away.

At your final stop, take off those sunglasses and look through one lens while rotating it. Do sections of the sky turn darker? Do reflections on car windows vanish? Then your glasses are polarized and we're set for more science. Reflections are everywhere in life. Except when you look right at the Sun (a no-no), all the light you see has bounced off something, and this bouncing tends to align its waves in one particular plane, like wheat rippling in unison along a field. Polarized lenses are transparent to light vibrating in one plane while blocking much of the rest, which is why you can now magically remove reflections from objects just by rotating the glasses. You'll also see colors and patterns on windshields that are simply invisible without your sunglasses.

Now that you're parked and walking to the office, you might notice that shorter people take faster strides. Ever wonder why? Everything has a natural resonance rate, the way your car's windshield resonated with light. A pendulum a yard long leisurely swings once a second but doubles its rate if it's half as long. Well, people's legs are something like pendulums. Without ever thinking about it, we let our limbs swing at a certain rate, a natural gait determined by the length of our legs. Of course, people can consciously choose to move faster or slower, but any pace other than their natural pendulum swing requires more effort. So the world is filled with walkers pacing according to the pendulum principle.

And your own strides finally carry you into the office, where someone greets you with "Hey, what's happening?"

Normally you'd say, "Nothing much."

But today you reply, "Refraction, reflection, diffraction, polarization, aberration, interference, Bernoulli's principle, the Doppler shift, Earth's motion, electron resonance, the energy-absorbing liquid-to-gas phase change, and the pendulum effect on walkers . . .

You?"

Chapter 3

We've Got Gas

"*What's that made of?*" The answer would have been self-evident not too long ago. Everything of human manufacture was either metallic, wood, ceramic, or stone. Now, however, we're overwhelmed by fiberglass, composite boards, and thousands of various plastics. No wonder we've all but given up trying to understand what comprises the stage sets in which we conduct our lives. When some unknown viscous substance clings to a finger, we may shriek, "*Yow!* What *is* that?" but we don't *really* want to know. Few of us even know why soap will get that stuff off. (Answer: Soap molecules have two open "hooks"; one end bonds with water, the other with oil. Each molecule grabs a bit of oily dirt while also attaching to the water that carries it away.)

One of the true compositional basics left is the air we breathe, the most intimate and communal aspect of our planet. As we take our next breath, we also inhale at least one atom exhaled by every person who ever lived. Or, more accurately, every deceased person and all living people older than the age of six, since it takes six years for an exhaled breath to thoroughly mix into Earth's atmosphere and have its atoms breathed in by everyone else on the planet.

And so it has been ever since our original ancestors abandoned gills for lungs. Yet it was only a century or so ago that two of our most brilliant minds managed to uncover the nature of the invisible gases that envelop and permeate us. Their discoveries have remained relatively unknown. A shame. The atmosphere may seem insubstantial, but it's more than our constant sustainer—it's filled with surprises.

From birth to death we inhale mostly nitrogen, which comprises 78 percent of our air. It doesn't hurt us and it doesn't help us—a little like elevator music.

Air's second component—one-fifth of it—is oxygen, high on everyone's favorite-element list. It's the only element that people can buy from dispensing machines (in places like Tokyo). Though the third most prevalent element in the universe, oxygen is particularly abundant here on Earth, where it has amassed itself in greatly condensed form, like concentrated orange juice. It is everywhere, having wormed its way into most surface rocks and two-thirds of each bit of sand and quartz.

Earth is the only known celestial body with an oxygen atmosphere. Oxygen would not even be expected as a major component of a planet's air, unless it had been created by plants as a by-product of photosynthesis. Because oxygen produces a green aurora when stimulated by stellar subatomic particles, any detection of green light around a distant world would indicate that life might be found on its surface. It's as if, throughout the universe, plants flag their green presence with an atmospheric glow of the same color.

So much for 99 percent of the atmosphere. What's the other 1 percent of the air we breathe? Put another way, after nitrogen and oxygen, what's the most common element you have inhaled, nonstop, since the moment you were born—the number-three substance that forever surrounds us?

Guess.

Many people would say carbon dioxide, but that's not an element, and anyway it comprises less than 0.001 of the atmosphere, despite its notoriety as a greenhouse gas. We assume that carbon dioxide is a major atmospheric player because we're inclined to pay attention to villains, and CO_2 gets unrelentingly bad press. Understandable. Carbon dioxide, which we exhale, is the gaseous equivalent of urine. We don't want it, and the fact that plants like it only diminishes its status—after all, they enjoy ammonia and horse manure, too. Carbon dioxide is not the third component of our air, and neither is water vapor, which, unlike everything else in the atmospheric brew, varies so greatly from place to place and hour to hour that it's not a reliable component.

Hydrogen? No way. There's over three times more hydrogen throughout the universe than everything else combined, but you wouldn't know it from what's doing here on Earth. In contrast to our oddball planet, places more compositionally characteristic of the cosmos (like Saturn, the Sun, or the vast beautiful nebulae) are overwhelmingly hydrogen, and after that you'll find helium. But not here. Whatever leftover earthly hydrogen didn't find a secure niche by linking up with other elements, like the H_2O of water or the CH_4 of methane, found a precarious perch high atop the atmosphere. There, random motion from the Sun's heat bounced the hydrogen atoms around like loose tennis balls until they leaked into space. By now the remaining free hydrogen has been reduced to bystander status in Earth's envelope.

Know the correct answer? Not many do. It's an element discovered only a century ago.

Argon.

This is the gas used to fill light bulbs, so we look at it often but think of it never. Which is only fair, since argon does nothing to facilitate our thinking process—unlike hydrogen and oxygen, the main stuff of which nature fashioned our brains.

Argon was discovered by a Scot, William Ramsay (1852–1916), who eventually won the Nobel prize in chemistry for his work with gases. He also discovered the other so-called noble gases: neon, krypton, and xenon. And it was Ramsay who sent electricity through inert gases to produce the brilliantly hued neon lights that now fill our nocturnal hours with nouns such as "Beer" and suggestions such as "Eat."

What a simple idea: High voltage excites gas trapped in a tube until each atom's electrons leap into a higher orbit, which they don't enjoy doing at all. So at the first opportunity (in a fraction of a second) they tumble back down again, emitting a particle of light of a precise color.

That was just a hundred years ago. Who's heard of Ramsay today? You'd think that someone who discovered more of the universe's ninety-two natural elements than anyone else, ever, and was solely responsible for our culture's neon surrealism would be remembered for at least a single century.

On that same island a century or so earlier, the English pastor Joseph Priestley was so impressed after meeting Benjamin Franklin in London that he started tinkering with science and ended up the other big star in the vaporous firmament.

First, Priestley discovered carbon dioxide; it was simply the common gas that extinguished flame. Then he devised a way to make it dissolve in water, thus inventing the carbonated beverage. For all his later achievements, this is what earned him the Royal Society's Copley Medal, which back then was even better than an Oscar. Priestley's seltzer and Ramsay's neon lights together would give us the first nightclub. In fact, Priestley later discovered nitrous oxide (laughing gas), which became all the rage recreationally in the late nineteenth century.

In 1774 the good pastor did something simple but clever. He sealed a burning candle in a jar containing a plant whose roots had a little water at the bottom, and of course the flame soon went out from lack of oxygen. But later he found he could light it again. The plant had *created* oxygen! So Priestley discovered both photosynthesis and oxygen simultaneously. Back then, such basic discoveries were there for the picking.

Unfortunately, there was no shortage of oxygen in Priestley's neighborhood, and his house quickly burned down when set aflame by locals miffed at his support for the French and American revolutions. He grabbed his family and emigrated to Pennsylvania, where he died in 1804. Forty-eight years later, Ramsay was born.

By William Ramsay's heyday, new gases weren't so readily floating around awaiting discovery. The bright colorful lines seen through a simple spectroscope when any element was excited by electricity had made substance identification foolproof. Even the Sun's entire complex pattern had been matched up with the spectroscopic signatures of various earthly elements—except for a few stubborn cryptic lines first seen during a total eclipse in 1865. Apparently some mysterious substance existed there in our nearest star but not here. It was named helium, after Helios, the Greek god of the Sun, and was the only element that seemed to live elsewhere in the universe but not on our planet. Then in 1895, helium, the universe's second most prevalent substance, was found on Earth by none other than Ramsay, who matched its spectral fingerprint with the Sun's remaining unidentified lines.

Too bad Ramsay can't survey our present world and see how his discoveries have changed things. What would he think about our neon jungle? No natural light—not starlight, moonlight, sunlight, meteors, rainbows, or lightning—emits light that way, which explains why neon lights (and fluorescents, which shine by the same process) feel so alien.

Just for the record, not all "neon" lights contain that gas; only the orange and red ones do. Most of the other colors in so-called neon lights are produced by tubes of colored glass containing—you guessed it—argon, mixed with a little mercury.

Argon and most other gases don't have much of a place when it comes to Earth's solid mass, and neither is argon a principal component of any other planet's atmosphere. The four giant planets of our solar system consist mostly of boring brews of hydrogen compounds with a bit of inert helium thrown in. Our two neighboring planets each offer little else but carbon dioxide—crushingly hot and thick on Venus, cold and thin on Mars. It takes 100-mph Martian gales to put enough *oomph* in that skimpy air to raise the planetwide dust storms that periodically envelop that world, erasing all surface features as if our telescope mirrors were suddenly steamed over. Thus Martian air reveals its presence, while all but the single atmosphere-free planet Mercury wear a gaseous shroud, armor against the hard vacuum that extends to the edges of spacetime.

But not here. Unlike all the other planets, our own atmosphere is always transparent. True, water vapor often condenses out to form clouds, but clouds are not gaseous, they are droplets of liquid. This sometimes opaque quality of H_2O, plus its feel against one's face on a foggy day, makes people think that steam, fog, or mist is heavier than the rest of the air, and thus the floating clouds seem illogical. But contrary to common sense, water vapor is lighter than every other major component of our atmosphere. Dry air is actually much denser than moist air. That's why airplanes require faster speeds and longer runways to take off in foggy weather. Counterintuitive as it may seem, humid air is thinner and offers less substance for wings to push against.

The transparency of our atmosphere's constituents helps explain argon's ability to hide from the human mind until a mere century ago. Meanwhile, the person responsible for our thoughts about it has vanished from cultural consciousness, his fame no more substantial than the gas inside his neon tubes. Let's salute him now, imagining his name blinking on and off in bright orange, like a beer sign: RAMSAY. RAMSAY. RAMSAY.

Chapter 4

The Man Who Fell to Earth

What else is as invisible as the air we breathe, as critical to our existence—and as taken for granted?

Gravity, that's what—one of the universe's four fundamental forces. Could anything possibly be fascinating about it? Yet its eternal presence turns out to be one of the most enigmatic aspects of the entire cosmos. The fact is, nobody can explain its nature. Albert Einstein couldn't, and neither can Stephen Hawking or any other science luminary today.

Let's watch what happens when we fall. Any kid who has jumped from a ledge knows that the greater the leap, the harder the landing. Why? Because the higher we are, the faster we're going when we hit the ground. Maybe you recall the formula for the acceleration of a falling body: *Thirty-two feet per second per second.* Schools insist on expressing speed that way (or as 9.8 meters per second squared), but it's much easier to grasp in everyday language: twenty-two miles per hour. Fall for a single second and you slam into the ground at twenty-two miles per hour. Each additional second you're airborne makes you hit the ground another twenty-two miles an hour faster. Fall for three seconds and it's a sixty-six-mile-per-hour impact. Simple.

To be airborne for exactly one second, you'd have to jump from a height of sixteen feet. That's about half as lofty as the intimidating topmost diving board at most Olympic-sized pools. One and a half stories tall. Painful, especially if you belly-flop, but rarely fatal.

Two seconds of falling, achieved by jumping off a five-story building, accelerates you to forty-four miles an hour, an impact usually not survivable. To determine how fast a marble would fall if tossed from the Golden Gate Bridge, or the speed a bungee jumper attains when leaping from a 200-foot perch, just whip out your pocket calculator. Multiply the height in feet by 64.4 and then hit the square-root button. That's the final velocity in feet per second. To express it in miles per hour, multiply again by 0.68, or just round off by two-thirds.

How fast would you plummet if you jumped from the highest possible place (that is, fell toward Earth from infinitely far away, even beyond the Moon)? Disregarding air resistance, you'd come in at a sizzling 25,000 miles an hour. That's exactly the same speed needed to go the other way, to escape Earth with a single upward blast, as fired from a cannon. So the velocity you need to free yourself from any celestial body is the same as the speed you'd crash into it after falling from an infinitely great height.

For the Moon, that's 5,368 mph. For Mars, it's 11,000 mph. For the Sun, it's more than a million miles an hour, or 384 miles per second, the speed at which an incompetent, drifting, out-of-fuel alien spacecraft captured by the Sun's gravity would be sucked into its gassy surface.

Here on Earth, air resistance quickly slows things down. Skydiving classes teach that if you spread your arms and legs and let your body present maximum surface to the air, you won't fall any faster than 120 miles an hour. That's called "terminal velocity" and it's reached after jumping from a height of just 500 feet, or fifty stories. Unless you assume a streamlined diving position (in which case you can bring yourself to nearly 200 miles an hour) you won't accelerate further by jumping from anything higher than the fiftieth floor. Those daredevils who leap from lofty rooftops usually choose much higher buildings, but that's to buy themselves sufficient airtime for their parachutes to open, an excellent idea.

For lightweight objects, terminal velocity can be very slow. The tiny water droplets that make up clouds have typical terminal velocities of just a half-inch per second—so slow they'll remain aloft given the slightest updraft. But as they collide with one another, the size of the droplets increases until their terminal speed exceeds the updrafts. Result: rain.

But why should Earth attract rain, or your body? And (ignoring air resistance) why shouldn't you be pulled to Earth more quickly than a lighter object like a pebble? The answer is: You are. Despite the apocryphal demonstration by Galileo from the Leaning Tower of Pisa—in which he is supposed to have dropped objects of two different weights to show that they would hit the ground together, thus proving that everything is pulled equally by gravity—heavier objects *are* yanked more forcefully than lighter ones. Your body *is* tugged with more force than a pebble.

If you jump off a table just as a fork falls off, Earth's gravity pulls on your body with greater force than it pulls on the fork. But since you weigh so much more (especially after dessert), your mass takes longer to speed up, just as a truck accelerates more sluggishly than a sports car. The result is a wash: Your body is yanked more forcefully but accelerates more reluctantly, and you end up falling at the same rate as the fork.

The answer to the most tormenting question—not why we can't seem to forgo dessert but why all this happens—eludes us. Newton's notion of a mysterious unseen force is no longer accepted. And Einstein's idea of gravity as an aspect of the warping of spacetime doesn't fully explain it either. Schoolchildren are still taught the older, Newtonian way—that Earth circles the Sun because of the Sun's gravitational "pull." Few school science curricula offer the superior Einstein concept—that Earth falls along a straight path through curved spacetime. Because the Sun warps the space in its vicinity, our planet merely follows the shortest, straightest, laziest path (called a geodesic) in this curved spacetime—a path that carries us back to our starting point after a year.

Likewise, shuttle astronauts do not float around because they have escaped Earth's gravity. There is nearly as much gravity 250 miles up as there is on Earth's surface. Orbiting astronauts feel weightless for the same reason skydivers do—because they are falling freely.

Visualizing gravity as a consequence of a region's three-dimensional geometry (four dimensions when you include time) allows us to calculate motion with wonderful accuracy. In sum, heavy items like the Sun measurably warp the local spacetime, and this curved space then dictates how everything in its vicinity (including the Sun itself) has to behave. It's a crisp, symbiotic relationship. Nonetheless, our ability to compute how gravity behaves still doesn't tell us what gravity truly is. Of the universe's four fundamental forces, gravity is the least understood. The other three—electromagnetism and the two nuclear forces, weak and strong, that operate only within atoms—have been theoretically linked. Physicists even use the term "electroweak" to designate the combined weak and electromagnetic forces, having provided compelling reasons why these two in particular should merge during fantastically hot and dense conditions such as during the first moments after the Big Bang. Gravity alone has defied all attempts to weave it into any larger picture, to connect it with the other forces of nature, to unify it with the totality of the cosmos.

Is its power dependent on the structure of the rest of the universe? Would the value of the gravitational constant (the universal quantitative relationship between mass and the gravitational field it produces) change over time, as clusters of galaxies move ever farther apart? Will Earth's gravity magically grow weaker as the universe expands?

Can gravity be the influence from some other dimension that partially spills over into ours? Is it spread by theorized particles called gravitons, whose existence has yet to be confirmed? Is a dropped object actually motionless, while the floor accelerates through spacetime toward it?

We may never know. But we'll keep working on it. The ancient conundrum of gravity will always be attractive.

Chapter 5

View from a Window Seat

There are more productive ways to spend your time aloft than worrying about gravity. Can a long business trip, say, be converted into a scientific adventure, an odyssey of curious discoveries? Absolutely. It involves little more than investing an extra half minute on the phone while making plane reservations.

Few seem to realize how many amazing experiences are available to the jet passenger. In these opening years of the third millennium, the greater wonder might not be the fact of flight but its ready accessibility. Yet the marvel seems to have lost its fascination. Most people you see in airports look chronically bored, unimpressed with the idea that they are about to do what 8,000 generations of their ancestors could not. At 40,000 feet there are phenomena that simply cannot be seen elsewhere. So instead of an airport paperback or an in-flight movie, consider a novel mode of entertainment. Open the window shade.

What an idea! Except that the blinding sun allows little to be seen but the wing. So let's start from scratch and plan the flight to accommodate the sights uniquely available from seven miles up. Some of the sky's finest gifts can be viewed only by those who have entered it.

A window seat, of course. You'll want it either in the very front or at the rear. In a 747, the wing can be avoided by choosing row 40 or beyond. In a 737, you'd want to stay away from rows 8 through 17. Choose the first ten rows or the final five in a 727, the final fifteen in an Airbus or 757, 767, or 777.

Next, you don't want the Sun on your side. If the flight is west to east, choose the left side of the plane—seat "A" in most aircraft. An east-west route dictates the right side. In a north-south flight, time of day becomes important; the Sun will be on the left only in the morning.

As soon as the jet completes its climb through the clouds (or during descent) look downward to find the glory. Seen only on cloud tops, the glory is a series of colorful concentric rings surrounding the spot on the cloud surface exactly opposite the Sun. That antisolar point requires no great search: It's marked by the plane's shadow. Moreover, the vivid rings of color are explicitly centered on the part of the shadow that corresponds to your seat! If you're in the tail, then the shadow of the plane's tail is the center of the glory.

Caused by diffraction among tiny water droplets, the glory can vary enormously in size as it floats like a phantom on the upper side of the cloud layer. This itself is curious, since most other rings and arcs, such as halos and rainbows, come in only one size (with a single additional, dimmer arc sometimes seen). The glory's dimensions depend on the droplets comprising the clouds below. The larger the glory, the smaller the drops of water.

Do not give in to the impulse to tell the person sitting next to you, "I see the glory!" But do contemplate what exactly is going on, and why you've never seen this phenomenon before. Remember that these halos are below you and centered on a point opposite the Sun. When, in ordinary earthbound life, does such a thing happen? On Earth, the spot opposite the Sun from your eyes will be on the ground, marked by the shadow of your head. Since sunlight always bounces back from that spot to some degree (which is just the way a movie screen works), you can often see a bright glow like a halo surrounding the shadow of your head on a dewy lawn, an apparition known as the *Heiligenschein.* But you need a cloud or a wall of fog in that position to get the diffraction effect of a glory. The glory's nearest relative on Earth is the Brocken bow, a phenomenon sometimes seen in the Alps but available from any mountaintop, given the unlikely circumstance that you are looking downward onto a nearby cloud that's also opposite the Sun. A jet, during the rather brief period when it is in sunlight just above the cloud tops, provides the right ingredients routinely. The glory's beauty and effortless availability make it a rewarding quarry. How sad that so few air travelers know about it!

As the aircraft climbs and its shadow on the clouds below shrinks, the glory fades, the sky darkens, and the horizon drops. In effect, you are increasingly peeking over the curve of Earth to ever greater distances. The Moon or Sun will appear even if it is beneath the horizon for people on the ground below, and a low Sun or Moon—that is, more than 90° from the zenith—will clearly float below you.

At full cruising height, you are treated to a miniature version of an astronaut's perspective: more than 180° of sky. Actually, astronauts have a more limited panorama than is commonly realized. They orbit no higher than 350 miles, or one twenty-fifth of our planet's diameter above the surface. Relatively speaking, they are barely above the Earth. The strongly curved horizon depicted in live TV shots from space are artifacts of the extreme wide-angle lenses of the space shuttle's cameras. To an orbiting astronaut, the horizon is very nearly as flat as our own, and Earth really does not seem to be a body floating in space. The images they televise, even when appearing to depict half of a globe, fail to show entire continents or any likeness to the true whole-hemisphere views of our planet seen en route to the Moon.

As you gaze out the airliner's window, how far can you see? The formula for figuring the distance to the horizon is so uncomplicated that it can be instantly memorized. It's as easy as 1-2-3: that is, 1.23 times the square root of your altitude in feet gives you the horizon's distance in miles. Calculating this is easier than it sounds. For example, say you're sitting on a beach chair at the water's edge, your eyes some four feet above sea level. The square root of 4 is 2, times 1.23 yields 2.46, and there you have it. The horizon is about two and a half miles away. No wonder it's such an obvious, sharp line out there at sea; it's surprisingly nearby. Any small object farther than 2.46 miles away is out of sight, over Earth's curve.

At 40,000 feet, the arithmetic is almost as simple. That number's square root is a gratifyingly even 200, times 1.23 becomes 246. So you're 246 miles from your horizon. This means that at cruising height you're centered on a visual circle that takes in a vast 500-mile stretch embracing some 190,000 square miles. That's about 7 percent of the entire contiguous United States. Always within view is the equivalent of Austria and Switzerland combined!

Come nightfall, you'll see luminous dots outside your window. Above 75 percent of the molecules that comprise Earth's atmosphere, the sky looks noticeably darker and there would be 30 percent more stars visible were it not for the slightly light-screening layers of glass and Plexiglas standing between you and breathing difficulties. If the Moon is visible, it will appear larger, but not because you're closer to it. Actually, unless it's more than halfway up the sky (in which case it becomes increasingly difficult to see from your seat) it's no nearer than when you were on the ground. Even if it were overhead, you'd only be closer to it by 1 part in 30,000. The Moon's larger size is an illusion caused by its being framed in the small window.

Gazing downward, you see the effects of a century's electrification. The bluish dots are mercury-vapor streetlamps, routinely used in the third world and still common in the United States, despite their energy inefficiency. The pinkish-yellow dots are the newer, sodium-vapor lights, whose current prevalence has transformed the aerial scene since the 1960s and is responsible for the amber glow above cities.

The jet engines' hum (generated by a relatively crude, inexpensive petroleum product that is essentially kerosene, which fills enormous tanks in the wings) can lull you into forgetting the passage of time. But contemplating time is a good way to spend it, observing how the interaction between moving jet and rotating planet alters the rhythm of day and night.

On the ground, the spinning Earth carries us at a rotational speed that varies with latitude. Got a calculator or love math? Then you can quickly find your current cruising speed on our whirling planet, for it's simply the cosine of your latitude multiplied by 1,038. But if words like "cosine" turn you off, just forget the whole thing; I've done the math for you:

People at the equator speed by at 1,038 miles per hour, while New Yorkers shuffle along a bit slower for a change, at about 785 mph. Penguins at the South Pole are essentially stationary, performing a single leisurely pivot every 23 hours, 56 minutes.

It's come to pirouetting penguins to make this point: Your jet, tooling along at some 550 miles an hour, can't quite nullify the Earth's rotation. That is, unless you're traveling at a greater latitude than 65° or so, such as on a polar route or passing east to west over Alaska. In that case, the Sun can indeed move backward in the sky, or set and then rise again in the west, as it does on Venus.

When you land in Australia or any other destination well into the Southern Hemisphere, the Moon and Sun *always* move backward. The Sun still rises in the east and sets in the west, but (as you face it) it travels left instead of right, unlike the way Americans, Russians, Europeans, Japanese, and most of the people of the world have seen it cross the sky all their lives. Tourists there often sense that something is deeply strange but cannot immediately say why. (And while we're on the subject, water does not spiral down drains differently when you're "down under." The Coriolis effect applies not to sinks and toilets but only to phenomena that cover significant distances. In the Northern Hemisphere, projectiles and air masses curve to the right when moving south or north because there's a difference in the rate of Earth's spin beneath them as they travel. By contrast, the difference in Earth's rotation beneath the perimeter of a basin or toilet bowl just about matches the speed of a wall clock's hour hand. Too small to influence the path of the water, here or there.)

When you fly the other way, west to east at high latitudes, the days and nights flash by in almost time-machine fashion. A full cycle of daylight and darkness can be compressed into less than twelve hours. All in all, the view from your seat is enlightening, but it's got tough competition. Which wins—observations and thought experiment, or the in-flight movie, *Terminator 8*?

Chapter 6

Fathoming Water

Flying back one winter night from a conference in New York City, our noisy little four-seater followed the river up the mid-Hudson Valley, treating us to a vivid demonstration of nature's skill at concocting countless creations from the same material. Thirty minutes after takeoff, the terrain changed to snowy squares of blue moonlit fields. The fog that had covered Brooklyn near the ocean gave way to clear air, and now we could see water in other forms: frozen ponds and lakes passing beneath the wings. Except for a strip cleared by icebreakers, the Hudson was solid as well. Reassuring. Every frozen body of water is a potential landing site in a pinch.

Although it looks inviting down below, water is a pilot's enemy. Thunderstorms, ice on the wings, water in the fuel lines—no part of water is welcome aloft even as the very brains formulating that aversion are themselves mostly composed of it. As we carefully avoided huge cumulus cauliflowers whose tops glowed brilliantly in the moonlight but whose moist dark interiors would have had us flying blind, there came an awareness of gliding like some interstellar drifter amid wildly different stuff all fashioned from a single substance.

That substance also happens to compose three-quarters of the universe. Plain hydrogen. What could be less interesting? Hydrogen, normally a yawn in chemistry lab, has managed to command the world's attention only a few times in the past century. When the *Hindenburg* blew up while attempting to land at the Lakehurst Naval Air Station in New Jersey in 1937 and the *Challenger* exploded soon after liftoff in January 1986, the fascinating and horrible spectacles illustrated the simplest possible chemistry. Here was hydrogen releasing itself from a manmade cage to find its way to its favorite companion, oxygen. Their eternal offspring is water, so that during the explosions the white billowing "smoke" surrounding both dying airships was nothing more than water vapor. Ordinary cloud.

In the early 1950s, hydrogen frightened us when TV showed the frequent mushroom clouds of H-bomb testing. That "H" of course was hydrogen, being transformed into helium in a thermonuclear fusion that converts bits of leftover mass into pure energy à la $e=mc^2$. The process has only a 0.7 percent efficiency—that is, only a small fraction of its material is changed to brute power—but that's still enough for a few pounds of hydrogen to destroy a city.

Sunlight is a direct manifestation of hydrogen's activity, and so are the rings of Saturn and the red nebulae so intricately airbrushed throughout our galaxy's spiral arms. We rarely think of hydrogen as being a dominant player, yet it manifests every which way throughout our lives. In a curling wisp of cloud drifting above, its patterns mutating like the letters of an alien language. In a line of blue icicles hanging menacingly from a rock ledge. In the boiling whirlpool of river rapids. As the major component of water, hydrogen plays its paramount role as far as terrestrial life is concerned. In a wheels-within-wheels shadowplay, hydrogen combines with oxygen to create a compound that assumes countless forms.

Our planet exemplifies nature's water motif. But not because Earth holds some unique status; the fact is, water is simply the most common compound in the cosmos. Not surprising. Hydrogen is the universe's most abundant element, and oxygen, though 1,000 times less prevalent, is the element that most loves to mate with others. Small wonder their courtship should be repeated in every corner of space and time.

Comets are essentially balls of ice, their glorious tails mostly water vapor. The moons of the five outermost planets are globes almost entirely of ice and water, nearly to their cores. Most of the night's stars, small, red and barely visible, are enveloped in gaseous water vapor—or steam, as it is known in our kitchens.

Freeze Frame

Water is commonplace, yes, but also very peculiar. When almost everything else in the universe changes from a liquid to a solid it contracts, growing denser in the process: Solid mercury sinks when placed upon liquid mercury; frozen butter sits at the bottom of a saucepan filled with cream. Water bucks this rule by expanding into a crystalline structure that takes up more space than the liquid form, thus becoming lighter, an oddity that is both a benefit and a major nuisance. If water behaved normally, our lives would be very different. Icebergs wouldn't float (the *Titanic* would have arrived in New York) and nobody could ice-skate except where a body of water had completely frozen from surface to bottom. Pipes wouldn't burst, either.

Even more than its oddly lightweight solid phase, water's freezing process is a curious one. This is because liquid water is heaviest at a specific temperature: 40° Fahrenheit (4° Celsius). Other liquids are densest just *before* they freeze. That water is densest 8° above its freezing point is not only extremely unusual but profoundly consequential. It explains why ponds do not freeze solid, allowing fish and other organisms to safely "winter over" near the bottom, and why large bodies of water like the Great Lakes are not ice-covered even when the air temperature lingers far below the freezing point.

Here's why. When a still-mild autumn lake is cooled by bitterly cold air just above it, its surface layer is of course chilled first, from 60° to 50° to 45° . . . colder and colder. But this top layer cannot possibly reach the 32° freezing point without first passing through 40°, and when it does it becomes denser than the water below it, and so it sinks to the bottom. Now there is a new surface layer, but it, too, can't freeze without first reaching 40° and sinking to the bottom.

In short, no freezing of the surface can take place until the entire lake's water, top to bottom, has reached forty degrees. Only then will the surface chill to 39°, 38°, 37°—all of which are less dense than the 40° water and therefore remain at the top. Finally this top layer freezes and the lake's ice thickens through winter from the surface downward, sparing the marine life that lurks safely below.

Since an entire lake's volume must be chilled to 40° before any surface freezing can occur, large bodies of water never reach that situation, especially since water is a poor conductor of heat. The world's major lakes thus have ice-free surfaces during most or even all of winter. See for yourself: If your love of science is strong enough, you'll surely want to dive in with a thermometer to prove it.

Another peculiarity of water is its property of melting sooner when under pressure, at temperatures lower than its freezing point. Called regelation, this process explains why kids can make snowballs. The snow sticks as a ball instead of simply falling apart as separate flakes because the hand pressure exerted while you're packing the snow abruptly lowers its melting point and lets the snow adhere. When pressure is released, the snowball instantly refreezes.

A good demonstration of regelation involves a simple block of ice and any thin wire, such as a guitar string. (The high E or B string works best.) Attach rocks to the wire's ends and let them hang to the side after placing the string atop the block. The pressure of the wire will reduce the melting point of the ice just beneath it, but the ice immediately refreezes as soon as the wire has passed down to the next layer. Result: The wire works its way through the entire block, yet when it's reached the bottom the ice isn't cut in two but remains solid and seamless. Magic.

Denver's Coffee

Isn't it convenient that we live on a planet whose average temperature hovers above that critical 32° mark much of the time? After all, ice exists for a generous 500-degree range (from absolute zero at –459.67° to +32°F) and steam manifests for nearly 3,000 degrees, starting at 212°F and going all the way to where atomic motion becomes so jittery that its molecular bonds break apart.

Liquid water inhabits an extremely narrow range of just 180 degrees, which happens to suit our environment and our bodies. It's only to be expected, then, that liquid is the state with which we are most comfortable and familiar, while ice and steam are the real templates for water's endless creative designs almost everywhere else in the universe. (Steam, incidentally, is invisible. The inch or two of clear air seen near a teapot's spout—that's the actual steam. The white mist farther from the spout, commonly mistaken for steam, is actually a spray of tiny droplets where the vapor has condensed back to its liquid form.)

But to have a waterworld, you need more than water and the right temperatures. Liquid water exists only under pressure. It needs to lie beneath a substantial atmosphere or other weight. Water, the bully that wears down mountains and tosses ocean liners, becomes a feeble invalid when air pressure is reduced. This effect has disheartened many mountain climbers, who watch their Coleman stoves appear supercharged as melted snow quickly boils long before reaching 212° or even getting very hot. Consequently, no café is likely to succeed atop Mount Everest; it's impossible to make a warm cup of coffee above 20,000 feet without a pressure device. Even in Denver, Colorado, a mere mile high, water boils at a temperature nine degrees lower than it does at sea level. Denver's coffee is always much cooler than Boston's.

The restaurant potential decreases further on worlds with skimpier atmospheres. On Mars, water evaporates so willingly that it boils into vapor and vanishes—if it doesn't freeze first. (At low pressures, both freezing and boiling happen at the same time!) You can't have liquid water on Mars, the Moon, or in space, even if you import it. This is why the many meandering fluvial channels discovered on the Red Planet by the *Viking* spacecraft in 1976 were so astonishing: They tell us not only that Mars once had copious water but that its atmosphere was much thicker than it is now. Might the small rocky world go through cycles where it takes turns being alternately friendly and hostile to life? Could creatures from long ago have burrowed beneath the Martian soil, where ice (and possibly even liquid water) still remain?

This pressure requirement also explains the existence of the first ocean discovered since the days of Balboa 500 years ago—the subsurface ocean recently detected on Jupiter's moon Europa by the *Galileo* spacecraft. Europa's ocean is possible only because it sits under pressure beneath a layer of ice some five miles thick.

The Water Network

But none of these oddities or otherworldly possibilities can hold a candle to water's simplest and least appreciated characteristic. It is this: The two hydrogen atoms chemically bonded to one oxygen atom that comprise water are linked not in a straight line but at an angle of 105°.

What's the big deal about that? Everything. All by itself, that seemingly textbook-dry fact has made life on Earth possible, and perhaps on endless other worlds as well. The 105° angle creates a kind of polarity, giving the oxygen a net negative charge and the hydrogen a positive one. These opposite charges create a network of weak but significant connections—so that instead of being a loose mixture of individual molecules, water is a lattice-work that behaves like a much bigger structure. This arrangement has tremendous significance. Without such hydrogen bonding, water would be like all the other molecules of its size and weight at room temperature—a gas. It explains why your veins are filled with fluid instead of vapor.

Why Do We Gasp?

Beyond composing most of our bodies, water may well have played a major role in our species' history. Some scientists are convinced that the genesis of the human race can be explained aquatically.

We didn't start out on the savannah as just a smarter variety of ape, goes this line of reasoning, originally proposed by Oxford professor Alistair Hardy a half-century ago and popularized in several of science writer Elaine Morgan's books (particularly *The Scars of Evolution*). Instead, our ancestors were stranded, probably in East Africa, during a period of rising sea levels. One large colony of apes became marooned and had to learn to live on the beach. Eventually they made their peace with the ocean and started using tools, because they needed to pry open clams, break open crabs, and otherwise bring home the bacon. Spending more time in the sea, they soon lost their furry coat. An aquatic lineage would explain our hair-lessness; like dolphins, whales, and other aquatic mammals, we shed our fur because it is an impediment in tropical brine.

Perhaps the hair on our heads remained so that our young could have something to hold onto when we swam. Our noses grew so that we could breathe more easily when trying to keep our heads up. Our fat became attached below the skin (just as it is in dolphins and whales) rather than as the separate layer characterizing all the other apes and land mammals.

When surprised or terrified, we gasp. Apes never gasp. Why do *we*? It doesn't make sense—unless it's the vestigial legacy of the breath we took before diving to avoid a predator.

And we do indeed (except for pilots) love water. Apes will cross water only when they must, or if there's food on an opposite bank. But they don't love it. We do: We vacation on lakesides and by the sea, and it's claimed that a newborn baby will not drown if submerged in water—while onlookers display that uniquely human gasp. Though the aquatic ape theory has yet to be widely accepted, schoolchildren a century from now may be taught precisely that explanation of our origins—an appropriate one in our watery universe.

If our lineage, lives, brains, food, recreation, and nearly everything else we care about involves water (which in turn is only one of hydrogen's designs), it's clear that the universe needn't harbor billions of disparate elements. There are variations enough from just two, hydrogen and oxygen. Throw in yet another, carbon, and you've now got complexity beyond comprehension.

Water, a basis for creative adventures here and probably on countless other solar systems, is a strange substance. But its myriad marvels typify nature's ever-changing designs, like the cloud patterns that forever capture the attention of children and lure aviators and dreamers to the sky.

Chapter 7

The Underworld

What lies beneath our feet?

When ancient Greek thinkers like Aristarchus observed Earth's curved shadow during eclipses and realized that our world must be a sphere, an issue arose that would not exist if our planet were flat. Every ball has an "inside." Well, what's inside ours?

Deep caves give the impression that Earth's interior is cool. Anyone visiting Howe Caverns in New York or Luray Caverns in northern Virginia, or any other cave complex—or anyone whose basement is below grade—knows that just a few feet beneath the surface the thermometer reads a chilly and consistent 52°F all year round. This subterranean chill suggested to the ancients that, like an ice-cream sandwich, Earth has a cold center.

But volcanoes and hot springs argue otherwise. Not to mention bizarre places like Centralia, Pennsylvania, where the ground is always blistering and smoke has been filling the air since 1961 because of an inextinguishable coal-vein fire, still burning despite a 40-million-dollar effort to put it out.

Mines provide some answers. After the consistently cool rock near the surface, temperatures start to climb and continue upward at an average of 1°F for every 100 to 200 feet of depth. Today's deepest working mine, a gold mine in South Africa descending to 12,500 feet, roasts at 131°F, with individual walls reaching 140°F. Only intense air conditioning allows the miners to do their job. Experimental borings have gone even deeper, to nearly 40,000 feet, which sounds impressive until you realize that it's only one five-hundredth of the way to Earth's center. Of human enterprises, what's deeper? Nothing.

The temperature at Earth's center is about 8,500°F, just 2,500° less hot than the Sun's surface. This fierce heat comes mostly from the decay of radioactive materials but is also left over from our planet's creation. Long considered by religions to be the location of Hell and by the party out of power to house congressional committees, Earth's interior nonetheless holds surprises, many only recently uncovered.

In the 1990s, researchers found that Earth's inner core—a sphere the size of Pluto—rotates faster than the rest of the planet. It's almost as if there were a planet within our planet. Just as unexpected, this 1,500-mile-wide inner core is solid. It's essentially one enormous iron crystal. An outer core of molten, electrically conductive iron some 1,500 miles thick surrounds this solid inner core; its rotation generates our magnetic field. Why the polarity of this field flips suddenly every few thousand years remains a mystery. Abruptly, *wham!*, the magnetic poles reverse themselves: North becomes south and vice versa, causing scout groups to suddenly wander aimlessly. Its timing is unpredictable.

The planet is extremely well constructed, and daily deforms only eight to eleven inches when tugged tidally by our large and nearby natural satellite. Earth's hefty overall density of 5.5 grams to the cubic centimeter, the greatest of any body in the solar system, is largely responsible for this reassuring stability. This heft in turn comes mostly from iron, Earth's most abundant element, and the primary component of the core and the mantle above it.

Floating on top of it all like pond scum is the uppermost region, the lithosphere, averaging just ninety miles thick. The lithosphere itself is cracked into some twenty major tectonic plates, which drift as much as four inches a year (in the case of the Pacific Plate, whose passage over a hot spot is in the process of creating the Hawaiian Islands) and whose interfaces delineate Earth's "ring of fire" regions of volcanoes and earthquakes. These in turn sport their own odd accoutrements, like the eerie and still controversial harbingers of ground tremors called earthquake lights. Many eyewitnesses have reported bizarre flashes or glows in the sky a day or two before or immediately preceding major quakes. Few have been documented scientifically. Nonetheless, there is a plausible explanation: the piezoelectric effect, in which masses of subsurface quartz become compressed by underground motion, releasing electricity. How this can find its expression in the sky is a puzzle, but at least we're halfway home in uncovering a mechanism by which nature could weave a light show linking subsurface earth and firmament.

Since we'll probably never penetrate deeper than the twenty-five-mile thick continental crust at the top of the lithosphere, its contents provide whatever earthly materials we have at our actual disposal. Care to guess what elements dominate? Nearly half the crustal inventory is simple oxygen, at 46 percent. About one-quarter (27 percent) is silicon. No surprise that their combination, as in sand or quartz, is so common around us. After these two elements, everything else is something of an afterthought. Aluminum accounts for 8 percent, iron 6 percent, calcium 4 percent, sodium 3 percent, and it goes downhill from there.

Standing barefoot on Earth's surface, you feel 5,000 times more heat coming at you from the Sun above than from the ground below. But on other worlds the situation is reversed. Jupiter and Saturn get twice as much heat from their own interiors as they do from the Sun. Since energy is needed for life, the prospect of life-forms existing deep below in low-rent districts of various worlds is starting to be taken seriously. In recent years, scientists have found living bacteria as deep as two miles below Earth's surface, in places forever deprived of even secondary solar energy. This suggests that seemingly lifeless worlds might teem with subsurface biology. As soon as technology allows, we'll undoubtedly send robotic probes capable of digging farther below the Martian surface than the foot-deep scrapes made by the two *Viking* landers in 1976. And—far more challenging—explore the oceans of Jupiter's moon Europa beneath their crust of sterile ice.

In common with chocolates, only some of the solar system's spheres have interesting centers. The Moon's core is cold. Jupiter's is almost starlike in its heat. That Jovian core is under such enormous pressure that its plain-vanilla hydrogen displays metallic properties and helps generate that planet's enormous magnetic field, the largest structure in the solar system. (Contrary to popular belief, Jupiter was never "almost a star." Even if it had fifty times more mass, temperatures would be insufficient to ignite nuclear fusion in its core.)

The most important core in our cosmic neighborhood belongs to the

Sun, where fierce nuclear-bomb conditions produce its light and heat. The by-product of the process is helium, steadily building up, some of which will eventually turn to carbon and later to iron, as if by alchemy. The cores of stars even more massive than the Sun become vast treasure chests of assorted elements that spew riches into the cosmos. Collapsing abruptly under their own weight, these huge stars undergo titanic supernova explosions that seed their environs with things like iron—which explains that element's ubiquitous and abundant presence on Earth and in our bodies. Indeed, every element that comprises our cells was originally created in a star's core (except for hydrogen, which is truly primordial, having begun forming shortly after the Big Bang itself). Even when the glitzy supernova show is over, the stage is not empty but occupied by a bizarre sphere of naked neutrons (see chapter 21).

Why spheres? Why are stars, planets, and major moons all globes and not cubes or dodecahedrons? For the Greeks, Earth's design fit beautifully with their philosophical belief that the sphere is nature's perfect shape. But science has its own hard-nosed reasons. When any massive object first forms, its own gravity causes it to contract into the most compact possible structure, having the property that no part of its surface is farther from the center than any other. Only a globe qualifies. Indeed, a sphere possesses the smallest surface area of any geometric shape. Low-mass objects, like asteroids and moonlets, don't shape up only because they lack sufficient gravity to pull themselves into balls.

Those remarkable MENSA-smart Greeks didn't know all these things, nor could they. Aristarchus and his friends were adept at using logic to solve mysteries of nature, but the planet's center was not knowable by deduction. Although Earth's core is nearer to Americans than Athens is, it's harder to reach than the stars.

The fun of finding buried treasure is far from over. Somewhere between core and surface, in layers still unknown, lie substances and perhaps even life-forms still to be discovered. No malevolent devils and damned souls, no underground *Matrix* city of Zion, no reclusive rock musicians. Something else, unimaginable, yet to come.

Chapter 8

Odd Odds

Vhat are the odds the Earth will swallow you up during an earthquake? It's a tricky business, measuring danger.

When applied correctly, statistics can fill hard drives with valuable if often counterintuitive information. But statistics, as we all know, are also often misused either deliberately or through ineptitude, and then all sorts of mischief bubbles up.

For example, government automobile safety stats for 2001 show that 39 percent of all auto accidents involve alcohol (a big improvement from ten years earlier). Still, if 39 percent of accidents are caused by drunks, the drivers in the other 61 percent are sober. In that case, shouldn't our first priority be to go after those responsible for most of the crashes and make a real effort to get the sober drivers off the road?

Statistics provide ideal material for stand-up comedy, as illustrated by this accurately quoted statement by a National Park Service spokesperson: "Too many people were killed last year in falls into the Grand Canyon." Too many killed? How many plunging people would have been optimal?

Though they capture most of the headlines, accidents are responsible for only 4 percent of all U.S. deaths during a typical year. Compared with heart disease (31 percent of deaths) and cancer (23 percent), all else is relatively minor. Yet the public's fear of accidents so skews the general perception that some accident insurance companies reap huge profits . . . with little risk. The odds are 25-to-1 that you'll die from something else.

In the United States, Canada, and Europe, accidents don't rank in the top five causes of death unless you're a teenager, and lamentably it's then number one. The odds of having an accident that damages your automobile but doesn't cause death are so high, however, that if you've driven for a couple of decades it's a near certainty that you've reshaped a car at least once.

But killing yourself in a car takes effort. The newest statistics peg fatalities at one per 75 million passenger miles. So a routine drive to the mall involving a thirty-mile round-trip carries a death risk of just one chance in 2 million. Those reassuring long odds tell us that driving is generally safe. (Here's a twist: While insurance premiums are usually lower for women than men, especially in the teen/young adult group, the overall automobile accident rate for females is consistently 15 percent higher than for men. Go figure.)

Yet even after a typical lifetime of driving, logging 15,000 miles a year for fifty years, your risk of eventually dying in a crash is just 1 in 88. Put another way, if you know eighty-eight people rather well (friends, close acquaintances, immediate and distant family members), odds are that one of them will die in a car crash sooner or later. But forty-four of them will probably succumb to either heart disease or stroke.

Mile for mile, motorcycles are eighteen times deadlier than cars. Small planes are about six times riskier, but that doesn't mean they're dangerous, given that cars are so safe. If you fly only in good weather, an hour's cruise in a small plane carries only a 1-in-100,000 fatality risk.

What's really rare is an accident caused by an "act of God," since the Deity doesn't appear to smite people with any great frequency. In the United States, lightning remains the most deadly of these natural perils. Even so, of the 2.5 million unfortunates who left us to our own stupidities in 2001, only ninety-five had been fatally jolted. And the odds of departing due to a bolt from the blue very much depend on your gender. Males are five times more likely to be hit by lightning than females. And not because Mother Nature favors her own sex, as she does with infant survival rates. It's because men and boys are more often in open areas fishing or playing golf. That's still no reason to give it all up for Ping-Pong; the lifetime risk of being struck by lightning (but probably not killed) is just 1 in 50,000. One way to improve your odds: Avoid golf or boating in lightning-prone Florida.

Accidents can be divided between the stay-at-home kind (fires, falls, choking on food, tripping over the dog, and so on) and those unscheduled departures experienced during travel, the vast majority inflicted by motor vehicles. Recalling the folk wisdom that most accidents happen at home, one might wonder if it really is safer to stay put. Answer: Stay put. All means of travel entail a greater risk than remaining in front of the TV. Of course, the peril depends on how far you roam.

On average, in an entire lifetime, each American travels about a million miles, a statistic unique to our time: Journeying that far would have been extremely unlikely even a century ago. True, a few extraordinary nine-teenth-century sailors or railroad conductors might have covered a million miles; most, however, would not live to tell the tale, since danger-per-mile was much greater back then.

In years to come, if and when our means of locomotion starts to approach the speed of light, things will be different—in many ways. Even a tedious hundred-mile-round-trip freeway commute would take only half a millisecond, but aside from the time savings something else would change as well. All distances in front of us would shrink dramatically. (For more on this, see chapter 30.) That distances aren't absolute but variable depending on one's speed is one of the most astonishing—and little realized—consequences of Einstein's special relativity theory. Length contracts at high velocities, so that the territory in front of you literally becomes compressed. At 99 percent of light speed, a seven-mile journey has physically become a single mile. The significance: You could cover great distances without the risks of traveling those distances, simply because length would have shrunk while en route so that you never actually traversed that much territory at all, despite arriving on the other side! You'd get there. And you'd beat the odds!

But again, statistics can rattle our assumptions:

The animal that causes the most fatalities? No, not shark, bear, rat-tlesnake, or spider. The lion's share of deaths is caused not by lions but by deer. Those cute Bambis are responsible for a hundred automobile fatalities in the United States annually and about $1 billion in damage. By contrast, the lifetime risk of your suffering a shark attack is 1 in 4 million. Alligators are twice as dangerous (1 in 2 million). Then come snakes (1 in 700,000), bears (1 in 410,000), and dogs (1 in 240,000). Deer may be charming but they're many times more lethal than all other animals combined—even deadlier than the figures quoted, which are attacks rather than fatalities.

Or consider the greatest weather hazard. Of tornado, hurricane, lightning, heat, cold, or flood, who'd guess that the highest peril comes from heat? Your lifetime risk of death from heat exposure: 1 in 10,000, more than all the others combined.

Birthdays matter, too, because accidents do not strike all age groups equally. Your peak years for dying from drowning are from birth to age five; for dying in a car accident it's ages fifteen to twenty-five; for succumbing to a poisoning it's thirty to forty; for dying in a fall, eighty to ninety.

The activity that carries the lowest risk while bestowing the greatest gain (reducing cardiovascular disease) is walking. So if you're really in the mood to play the odds, the best bet is: Take a hike.

People are equally clueless about statistics when it comes to potential gain. Since any activity that alters the odds of an occurrence by 1 in 20 million is essentially pointless, it makes no sense to participate in a state's lottery. Practically speaking, you don't increase your chance of winning by buying a ticket. Aptly, the lottery has been described as a tax on people who are bad at math.

Impossible!

Musing about such minuscule probabilities, it's easy to cross the border into speculation about what, if anything, is altogether impossible. Although we occasionally base decisions on far-fetched prospects, few of us want to spin our wheels on Futility Boulevard. But how unlikely must a phenomenon or event be before it can be safely dismissed as impossible? Is it possible that, out of human earshot, squirrels communicate with each other in French? Obviously we have to draw an arbitrary line at some point, since proving the negative is itself impossible. Still, science is based on observation, verifiable laws and phenomena, and hard evidence. That's why when someone on TV claims to have been taken into a flying saucer for medical experiments, we're more apt to say "Yeah, right!" than "I hope that was covered by their insurance."

On the other hand, quantum mechanics revolves around uncertainties and probability. These laws, which operate primarily at the subatomic level, state that things exist not in real places but only as possibilities and likelihoods. When enough time passes (even if that entails many billions of years) extremely improbable events *must* occur. Since everything is made of subatomic particles, this manifestation of the unlikely will eventually happen on the visible, macroscopic level as well.

Given sufficient time, things that are merely possible, no matter how remote, will chip away at the fabric of the universe and change the face of reality. For example, carbon-14, inhaled into our lungs with every breath, is a very stable isotope of carbon, but occasionally a bit of its nucleus will move to an "unlikely" position, accompanied by a tiny emission of energy that changes it into an atom of nitrogen. The laws of chance tell us that this rarely happens. Yet by the time 5,730 years have passed, half of a given sample of carbon-14 has indeed changed to nitrogen, a statistical process that yields such a high predictability that radiocarbon dating has become the standard for determining the age of ancient artifacts.

(Here's how the dating works. Cosmic rays—subatomic particles zipping through the universe after being created in violent places like supernovae—strike some of our atmospheric nitrogen and convert it to C^{14}. Like normal carbon-12, C^{14} combines with oxygen to form carbon dioxide, which is absorbed by plants and in turn eaten by animals. Every cell of every plant and animal therefore winds up with a little C^{14}, comprising about a millionth of 1 percent of their body's carbon. When the plant or animal dies, the C^{14} is no longer replenished and it slowly decays back into nitrogen. Therefore cotton fabric, bone, anything that was once alive, can be dated to an accuracy of 15 percent by determining how much C^{14} remains.)

At any given moment in the here-and-now, the statistical ambiguity surrounding the most basic building blocks of matter severely restricts our knowledge about individual subatomic particles, limits our certainty about atom-sized particles, but only very slightly reduces our confidence about larger objects. Still, it is possible, not impossible, that the White House will suddenly vanish and rematerialize in, say, Disney World. However, the amount of time required before it becomes realistically likely that any particular object with that much mass will change position would greatly exceed the life of the solar system.

This quivering in the ultimate stability of matter and energy doesn't weaken the solidity of physical laws. It simply tells us that we are surrounded by probabilities rather than certainties. It also means that Einstein was wrong. God *does* play dice.

Birthdays and the Monty Hall Problem

The laws of probability are not always kind to casual logic; they can be mad-deningly counterintuitive. For example, in any gathering of people, what are the odds that two individuals share the same birthday? With 365 days in a year, you'd think you'd need half of that (about 183 people) to provide a fifty-fifty chance of a shared birthday. But you need only twenty-three people. Try it at your next party or gathering: Ask each person to announce their birthday in turn. With more than twenty-eight participants, a match is practically guar-anteed—and invariably produces amazement.

Ready for a mind-bender? This one arose in the mid-1990s and tor-mented mathaholics throughout North America. It came to be known as the Monty Hall Problem.

The set-up is a standard game show in which the contestant is trying to win a prize secreted behind one of three closed doors. Behind two of the doors is an amusing but essentially worthless prize, like a sheep, while the third conceals something desirable, like a Mercedes. When the contestant chooses a door, the game-show host quickly opens one of the two remain-ing doors and reveals a booby prize. The host then offers the contestant a choice: Either stay with the original pick or switch to the other closed door.

That's the situation. The question: Is it in the contestant's best interest to switch, or does it not matter?

There are two very different and equally plausible ways to consider this problem.

> 1. At this moment two closed doors remain, the coveted prize is behind one of them, and therefore the chance of either door being correct is fifty-fifty. So it does not matter whether the con-testant switches. Since it's simple guesswork, one door is as good as another. This seems self-evident. Right?

Not so fast. Consider the alternative reasoning:

> 2. When the contestant originally made her choice, she had a one-in-three chance of guessing the right door, thus there was a two-in-three chance that the prize was behind one of the other

doors. When the host opened a booby-prize door, the two-in-three probability of success went entirely to the unopened door that the contestant did *not* choose. There's still only a one-in-three chance that the original selection is correct. Conclusion: The contestant doubles her chance of winning by switching at this point.

Both (1) and (2) seem examples of irrefutable logic. Yet they reach very different conclusions, and only one can be right. When computer programs mimicked this situation, the result showed that logic number (2) is correct. The contestant will win the prize twice as often by switching. In short, while the laws of probability may initially seem ambiguous or even contradictory, they are actually as clear-cut and inexorable as (given enough time) losing in Las Vegas.

Errors and Mischief

In applying statistics to daily life we need to guard against the tendency to assign probability after an event has occurred. This basic mistake is seen frequently in pseudoscientific or some "New Age" presentations, and has been exploited by demagogues throughout history.

To illustrate how this works, assume you've just watched three women wearing red hats cross a street while the Sun emerged from a cloud and a distant ambulance siren wailed. You could maintain that the odds of witnessing this exact sequence are millions to one against. Did you therefore see something very improbable?

Not at all. Once an event has happened, its chance of occurring becomes 100 percent. It is no longer unlikely in the least. Only a prediction made in advance can assign meaningful odds.

Say an astrologer wants to "prove" that planets influence one's future career. Poring over lists of planet configurations, famous peoples' occupations, and celebrity birthdays, and fishing for something out of the ordinary—he finds it. Then he offers statistical proof that renowned pianists are disproportionately Scorpios. This sort of thing is often presented in pseudoscientific publications, but trained statisticians quickly spot the mistake: One cannot legitimately ascribe correlations after the fact. It may seem valid, but it just doesn't work that way.

Unusual weather is one arena in which this error often pops up, since odd periods of heat, cold, snow, or rain bring out the expert in all of us. And of course people see unusual conditions as portents of the Earth's demise or at least a permanent change in climate. Researchers, however, find it much more difficult to tell whether anything odd has actually occurred. The problem with weather is that the unusual is normal. True, there is "hotter than average" weather, or cooler, or drier, or wetter. But chances are that one of those anomalies will befall us each season.

A good analogy involves flipping a coin twice. Getting two heads is unlikely, and so is getting two tails, or first a head and then a tail, or the other way around. But while a particular outcome is unlikely, the odds are 100 percent that one of these sequences must occur. In short, an improbable event must happen every time a coin twice falls to the floor. That said, can we claim that the outcome, whatever it is, is in any way unlikely? Predicted ahead of time, yes. After the fact, no.

We'll often find ourselves in the midst of some type of unusual condition. Floods, drought, extreme heat or cold are a daily certainty somewhere in the world. Viewed with this perspective, we find ourselves in a universe where everything is fascinating—and nothing is surprising.

Except, as described in chapter 1, a meteor that crashes through your roof and rolls under the kitchen table. Especially when the previous shooting star to strike a house in North America did so in the same town. Now, what are the odds of *that*?

· Chapter 9 ·

Measuring Madness

On September 23, 1999, an accident happened that may have set the record for astonishing stupidity. On that date, the $125-million Mars Climate Orbiter plunged into the planet's atmosphere—and promptly disintegrated. The cause? One NASA team had programmed the spacecraft's engines with English units (feet and pounds) while another had used a metric scheme. It was yet one more costly penalty paid by the United States for its perverse use of two different measurement systems.

Locking the barn door after the loss, the science community renewed its insistence that only metric units be used. Since the 1964 National Bureau of Standards' policy declaration that the United States would henceforth employ only metric and, later, President Jimmy Carter's push for national metrification, not much has changed. Techies deal in centimeters, while everyday life chugs along inch by inch.

The metric system uses convenient multiples of ten. It is sensible and nearly universal. But the U.S. Customary System is sometimes superior to the scientifically preferred expressions, and there aren't always equivalents. In everyday usage there's no real metric substitute for cup or pint. We order a "shot" or "jigger" of whisky and, at least when sober, are not even close to asking the bartender for a few centiliters. And in back alleyways little known to the public, strange units—like the measurement of the heat of chili peppers—still reign, having no conceivable metric equivalent. (In this case the Scoville Heat Index, which quantifies the amount of blister-delivering capsaicin in any particular pepper. All-time winner: the Habanero, at up to 577,000 SCI, compared with a mere 5,000 SCI for your average jalapeno.) In a myriad of customs, from gun barrels (.45 caliber is 0.45 of an inch) to recipes calling for tablespoons, Americans are not remotely approaching full-scale metrification.

Acknowledging this fact, major U.S. media and books and magazines concerned with science for the general public almost exclusively use miles and pounds—to the discomfiture of scientists, who blame them for perpetuating antiquated units. The Mars Climate Orbiter mishap spotlighted this ongoing contentious issue, as did an amusing historical note circulated on the Web a few years ago. It seems that the space shuttle's booster rockets were designed to fit on railroad cars so that they could be transported from their manufacturer in Utah. The rockets' width was therefore constrained by America's railroad gauge of four feet, eight and a half inches. This odd measurement arose simply because it exactly duplicated the British rail gauge, which originated with English streetcars—employed because existing roads had ruts that far apart, made by horse-drawn carriages. The roads had all been constructed with that axle-width so that the carriage wheels would fit within the ruts, which in turn traced their origins to the Romans, who built the first English roads. Purportedly, the Romans had standardized this breadth for their chariots because the rumps of two horses could fit into it without their hooves kicking the wheels. The point is that even our high-tech rocket designs are dictated by such ancient measurements, revolving around such basics as horse derrieres.

A hodgepodge of "standards" has been bequeathed to us by our forebears. Some—like the term "one cannon shot," which once meant three miles—are totally out of use. Yet this unit and its implications explains why U.S. territorial waters were originally set at that distance.

A surprising number of holdovers are still widely employed. For example . . .

Five and a half yards is one rod, a Saxon measure well established by the eighth century, which may have been based on the length of the pole used to control a team of yoked oxen. Forty rods equals one furlong, the length of a typical furrow. Exactly eight furlongs make a mile, as bettors watching their horse tire at the final pole are still unhappily reminded. Three miles equals one league, a term resurrected by Jules Verne and later by the Disney movie of his *20,000 Leagues Under the Sea.* Few seemed to notice that it would not be easy to descend 60,000 miles beneath the ocean's surface, or (unless you're Captain Nemo) to laterally travel that far while submerged.

Much more logical were units based on the 180°/360° motif used in geometry. Any angle 180° from the first describes the opposite direction, so it made sense to early cartographers to assign a longitude of 180° to a place on Earth as remote as possible from the 0° starting point, the prime meridian in Greenwich, England.

For the same reason, eighteenth-century German physicist Daniel Fahrenheit deliberately divided the span on his temperature scale between water's freezing point and its "opposite," the boiling point at sea level, into 180 degrees (from 32° to 212°).

In navigation, each of those 180 degrees east or west of Greenwich is further divided into 60 minutes of arc, which measure 6,080 feet apiece at the equator (officially 6076.11549 feet). This international unit is termed a nautical mile, and the speed expressed in nautical miles per hour is called knots. The unit—first measured by watching a moving ship pass evenly spaced rope knots—remains to this day the only officially sanctioned way to express speed in aviation and marine commerce and is even used on the returning space shuttle.

The foot was originally 1/6,000 of a minute of latitude as measured in England, and the number 6,000 (as well as 360 for degrees and 60 for minutes and seconds in time, navigation, and geometry) was chosen for ease of dividing, having more factors than multiples of ten. Indeed, 60 is the lowest number with a dozen factors, being divisible by 1, 2, 3, 4, 5, 6, 10, 12, 15, 20, 30, and itself and thus special since the time of the ancient Greeks. So, up to this point, many units are shown to have had a logical rather than a whimsical basis.

Volume expression is a different barrel of monkeys. Here we see a cornucopia of disparate units, from the "cord" of firewood (126 cubic feet) to the "board foot" (a volume of wood 1' x 1' x 1"). But real fun comes in gallons. All manner of containers were called gallons as recently as the nineteenth century. You had the wine gallon (231 cubic inches), the ale gallon (282 cubic inches), and the corn gallon, which alone equaled the traditional, fifteenth-century Winchester gallon (192 cubic inches).

In 1824, England finally ended this mess with its Weights and Measures Act, which declared that only the new imperial gallon would henceforth be legal. Replacing all others, it was precisely defined as a size that would hold ten pounds of distilled water at a temperature of 62°F. It equals 4.54609 liters.

Simultaneously, however, the old wine gallon that had just been abandoned by the British became the new official U.S. gallon. Equaling 3.785411 liters, it's still the standard today in the nonmetric world.

Depending on how you spend your time, you're likely to use various old units that are very much alive. Writing paper comes in reams (500 sheets, since 25 sheets make a quire and 20 quires make one ream). Two reams are a bundle (1,000 sheets), and 5 bundles make a bale. In the cotton industry, a bolt equals 40 yards while a skein is 3 bolts. Less well known to the general public, 7 skeins make a hank and 18 hanks make a spindle.

Do you like herring? Three or four of them (depending on the country) are sometimes called a "warp" of herring, while 33 warps make a long hundred. Long indeed—you get 132 fish (or only 124, in some circles). If you *really* want to throw a party, order a last of herring. You'd receive at least 13,200.

Trying to define "barrel" gets even fishier. In England, a barrel holds 36 imperial gallons, but in the United States a barrel contains only 31.5 gallons, even though American gallons are smaller. Beer barrels officially hold only 31 gallons, while U.S. proof barrels hold 40. In the petroleum industry, a barrel officially contains 42 gallons of oil.

Competing against such entrenched but confusing expressions, the metric system was first established two centuries ago in France. Its very name derives from its fundamental standard—the meter, which currently is neither logical nor meaningful. Originally it was supposed to represent one ten-millionth the distance between the equator and either pole. Then, when that was found to be wrong (due to early surveying errors), the meter was redefined by a particular bar of platinum in a vault near Paris marked by two precise scratches. In 1960 the meter was again redefined, this time as 1,650,763.73 wavelengths of the orange light emitted by krypton-86— under certain conditions. How's *that* for a household concept? In 1983 it changed again to the present standard, the distance traveled by light in a vacuum for 1/299,792,458 of a second.

A more rational foundation unit would be nice, like the distance anything falls in Earth's gravity in a round-number period of time. Or maybe how far light travels in a billionth of a second. (On second thought, we already have that. By coincidence, that's one foot!)

Even in the sciences, metric units are far from ubiquitous; inconsistency routinely exacts its kilogram of flesh. The aviation community, as high-tech as any, insists that altitudes throughout the world be expressed in feet and speed in knots. Astronomers use twenty-four-hour, sixty-increment "right ascension" to specify east-west position on the sky, but ninety degrees to express "declination," or north-south demarcation. Metric measurement, officially termed the SI (for Système Internationale d'Unités), is dumped overboard in jewelry ("carats"), gardening ("flats"), stocks ("points")—and wherever else people prefer the simplicity of the status quo.

The issue of scientifically correct terms goes beyond the Customary versus SI conflict. Specialists, in common with yesterday's high priests, insist on keeping laypeople as baffled as possible, as when physicians specify dosage using the Latin "TID" (*ter in die*) when they could just as easily write "3xD" for "three times daily." Maybe that willful professional opacity helps explain why astronomers have replaced light-years with parsecs.

What a bewildering thing to do! A parsec is 3.26 light-years, the nearly useless distance at which a celestial object would display a parallax shift of one arcsecond as Earth orbits the Sun. How many people can relate to that? Nobody would want to pigeonhole the entire universe in parallax terms except those in the socially isolating field of astrometrics. And since nothing exists one parsec away from Earth (it's much farther than the most distant planets but closer than the nearest star), why make it the standard for expressing all celestial distance? Especially when only the nearest stars, those within the closest one-billionth of 1 percent of the universe's volume, display any sort of measurable parallax to begin with. Obscurity glorified! The erstwhile light-year was a wonderfully logical unit, using the amazing and unique constancy of light speed. It is meaningful across the entire cosmos, and easily graspable by everyone. If the star Vega is twenty-five light-years away, its image has traveled that long to get here. We see Vega as it was twenty-five years ago. Simple. Intuitive. And that's saying a lot, considering that the unit expresses the inherently mind-boggling vastness of space.

But, alas, "light-year" may have been insufficiently obscure to please the higher reaches of the astronomical community. In the current literature, Vega's distance from us is often pegged at 7.67 parsecs. Nearby galaxies are specified in megaparsecs. And it gets worse. Nowadays even metric measurements that were *de rigueur* a few short years ago are jettisoned faster than yesterday's bell-bottoms. Grams and kilograms? Gone! Now the official unit is the Newton, which in turn amounts to joules per meter. If that doesn't slide easily from your lips because you can't tell a joule from a jewel, you probably have lots of company.

Oddly enough, the number of inches in a mile is essentially the same as the number of astronomical units (Earth-Sun distances) in a light-year; both are very nearly 64,000. This facilitates scale models. If the Earth were one inch from the Sun (in which case, our planet would be germ-sized and the Sun the dot over this "i") then the nearest star (Proxima Centauri, 4.3 light-years away) would be another dot 4.3 miles away. The nearest spiral galaxy (Andromeda), which is 2.2 million light-years away, would lie 2.2 million miles away, four times farther than the Moon. In other words, a scale model with a germ-sized Earth, though easily created and grasped, itself quickly assumes astronomical proportions.

People in the sciences would like the world to adopt a unified system of weights and measures, and metric is close to becoming that universal structure. Only the United States and a handful of small nations still refuse to metrify. But while a global system would be best, iconoclasts continue to point out that measurement units should be based on human experience, inasmuch as they are designed for human use. And here is where the Customary System holds an intuitive advantage. A cup was originally (and still is) about the amount of liquid one can hold in cupped hands. A foot is roughly the length of a man's foot, and that characteristic of being the distance light travels in a nanosecond would seem to make it a logical choice for a basic unit. An inch is about the span between a pinky's two knuckles. A yard is a long stride. Few metric measurements are similarly nature-related, but no matter. We'll all go metric eventually.

Until then, here's something that would have been commonplace only a quarter century ago but now seems a rather strange conversion table. It's increasingly hard to find: conversions from old-style Customary units we still commonly use to *other* old-style Customary units we still commonly use! They're presented not to help perpetuate the fading system but to open windows to let new light fall on small wonders.

Take a river's flow of water, which is sometimes stated in "millions of cubic feet per day." How can anyone visualize that? The conversion to gallons per minute makes the flow-rate somewhat more meaningful. Similarly, when you see that one "atmosphere" (an archaic term) is equal to 29.92 "inches of mercury" (also archaic) or 33.90 "feet of water" (ditto), it starts the mind going. You're almost forced to picture the weight of air above us pushing down on everything around us, pushing so hard on the surface of a dishful of water or mercury that it will drive the mercury two and a half feet up a hollow tube standing in the dish. And send a column of water (much lighter than the mercury) as high as a three-story building. Suddenly, you understand why you can siphon water, as long as you're not requiring it to flow upward more than three stories (or, to be exact, 33.9 feet). All this, at a glance.

FADING UNITS

TO CONVERT	INTO	MULTIPLY BY:
Atmospheres	pounds/sq inch	14.70
Atmospheres	inches of mercury	29.92
Atmospheres	feet of water	33.90
Cubic feet	cubic inches	1,728.0
Cubic feet	gallons	7.48052
Cubic feet	quarts	29.92
Cubic feet/min	gallons/sec	0.1247
Cubic feet/min	pounds of water/min	62.43
Cubic feet/sec	million gals/day	0.646317
Cubic feet/sec	gallons/min	448.831
Cubic yards	cubic inches	46,656.0
Cubic yards	gallons	202.0
Cubic yards	quarts	807.9
Days	seconds	86,400
Feet/second	miles/hr	0.6818
Gallons of water	pounds of water	8.3453
Gallons/min	cu ft/hr	8.0208
Gallons (imperial)	gals (U.S.)	1.20095
Gallons (U.S.)	gals (Imp)	0.83267
Horsepower	watts	745.7
Horsepower	foot-lbs/min	33,000
Inches of mercury	pounds/sq foot	70.73

TO CONVERT	INTO	MULTIPLY BY:
Inches of mercury	feet of water	1.133
Kilowatts	horsepower	1.341
Kilowatt-hrs	Btu	3,413
Kilowatt-hrs	Pounds of water	22.75
Miles (statute)	feet	5,280
Miles/hr	feet/min	88
Miles/hr	feet/sec	1.467
Miles/hr	knots	0.8684
Miles/hr	meters/min	26.82
Million gals/day	cu ft/sec	1.54723
Ounces (fluid)	cubic inches	1.805
Parts/million	lbs/million gals	8.345
Pounds/cubic inch	lbs/cu foot	1,728.
Pounds/sq foot	feet of water	0.01602
Pounds/sq inch	atmospheres	0.06804
Pounds/sq inch	feet of water	2.307
Pounds/sq inch	inches of mercury	2.036
Pounds/sq inch	lbs/sq ft	144.0
Quarts (liq)	cu inches	57.75
Quarts	cu yards	0.001238
Square feet	acres	0.00002296
Square feet	sq inches	144
Square feet	sq yards	0.1111
Square inches	sq ft	0.006944
Tons of water/day	gals/min	0.16643
Tons of water/day	cu ft/hr	1.3349
Watts	Btu/hr	3.413
Watts	ergs/sec	0.000001
Watts	horsepower	0.001341
Watt-hours	foot-pounds	2,656

By the dawn of the next century, we may all be thinking comfortably in meters. Until then we may as well just relax, pour ourselves a deciliter of wine, and enjoy the whole nine yards.

Chapter 10

Say What?

We're so accustomed to language that other ways of exchanging information generally escape our attention. But it's now clear that various animal species have no trouble getting their messages across. Glimpsing some of their communication methods may help us connect with other life-forms here on Earth—or even with aliens, should the need ever arise.

Communication of any kind requires a transfer of energy. There's no known way around it. Whether disturbance in air (sound, speech) or electromagnetic radiation (light, image, "signing") an energy delivery must occur, which constrains the expression of information to the speed of light, at best.

Some people believe in telepathy, whose speed is unknown. Although science has been unable to support or confirm it in any way, it may be an error to dismiss telepathy as we dismiss divination, palmistry, and other such nonsense. Its supposedly spontaneous nature may preclude laboratory duplication. (Witnessing many compelling instances has convinced me to at least keep an open mind. For instance, close friends of mine who are identical twins, both skeptical and no-nonsense men with masters and doctorates in the sciences, report that one would often find a tune running through his head and then hear his brother suddenly begin to hum the same tune at precisely the same point in the melody, and in the same key! Frequent occurrences of such highly unlikely events make the "coincidence" explanation untenable.)

Science has no understanding of how brains might transmit and receive in this fashion. Yet there are communication modes that are nearly as bewildering to us but which have been scientifically demonstrated.

Humans rely on symbolism. The word "tree," for example, conveys a categorical, general idea of a particular type of vegetation. But observing a specific tree involves thousands of perceptions, too numerous to be expressed symbolically: The subtle mottling of bark, patterns of light and shading, the motions of myriad leaves, the complex branching—there's simply no way to convey the actual living image with words even if one spent a year in the attempt.

Other life-forms get around this limitation by employing more direct methods. While humans rely primarily on the sense of sight, dolphins, in a milieu that is often murky, use sound as a primary sense. They emit intricate noises that echo like radar from the target, whether an enemy whale or a school of potential food-fish, producing a kind of detailed sonogram. Because sound is reflected in complex patterns by tissues of different densities, porpoises can even detect the internal structures of whatever creature they're "scanning."

But here's where it goes almost beyond belief. Dolphins also have the ability to reproduce the radar images that their ears perceive. So they swim to their friends and tell them what they saw. But they don't use symbolic speech. Instead they duplicate the echoes they picked up, directly "painting" the sonogram of the object they perceived into the minds of other dolphins—perhaps even emphasizing features of interest. In short, their friends can actually "see" whatever the first dolphin saw.

When we try to communicate with animals, an interesting two-way exchange usually develops. We still use verbal abstraction—saying "Stop that!" when Fluffy rips up the garbage—while the animal just as effectively tells us what it wants with nonsymbolic sounds. Pet owners quickly learn to distinguish from among dozens of cat or dog vocalizations. The canine's growls, whimpers, barks, and yelps accurately convey a full spectrum of needs, comments, and emotions, even if no single dog-word has a direct English equivalent. Obviously, "woof" has no meaning out of context.

Some assume that animals are dumb, that they don't "get it." But it's obvious to trainers, farmers, and pet owners that animals grasp everything that interests them.

The less a creature resembles a mammal, the stranger and more indecipherable to us is its method of information sharing. It's reasonable to assume, then, that interfacing may be a real challenge if ever we encounter extraterrestrials. Since they have assumedly arrived on some sort of spacecraft, they have obviously developed technology (as opposed to whales, say, who are generally regarded to be as smart as *Homo sapiens* but who have no such inclination). So techno-babble might give us something in common. If aliens are accustomed to any sort of symbolic communication (which is what sci-fi movies always assume), the task might be doable.

It's unlikely that newly arrived aliens would engage in the verbal formalities popular in fiction ("Greetings, human"), so superficial they might as well have mailed us a card and saved themselves the trip. Nor are they apt to enter into a silly musical exchange, as in the 1977 Spielberg film *Close Encounters of the Third Kind,* in which sophisticated aliens in a huge hyper-tech spacecraft deliver a musical composition a rat might create by running across a keyboard. (It's all part of the standard "I'm OK, you're an idiot" theme favored by the movies. Extraterrestrials either possess a sub-parakeet command of English, or they're taciturn because conversation is pointless when you intend to vaporize your audience.) In the sciences, it's hopefully assumed that math and basic physical laws or the simple geography of our region of space might provide the material with which to bootstrap our way to a common language. For example, every intelligent alien would presumably know that hydrogen is the most common atom in the cosmos. (But, alas, do most humans know this?) It's also assumed that they'd recognize a sketch of that atom with its single electron, so that hydrogen's size could be considered a basic unit of length, as could that element's characteristic radio emission at 21 centimeters.

A simple scale drawing of the solar system could introduce another unit, the Sun-Earth span. Pi, the ratio of a circle's circumference to its diameter, would certainly be known throughout the universe. A lingua franca for space-faring species could presumably be built with such units as stepping stones. But only if they were willing or able to use symbolism.

That was the assumption behind the messages accompanying the twin *Pioneer* and twin *Voyager* spacecraft, launched between 1972 and 1977 and now traveling in interstellar space. In *Voyager's* case, these messages are in the form of gold-plated video disks in protective sleeves containing images of cities, farms, people, animals, sunsets, and music ranging from Bach to Chuck Berry, with the sender's address schematically depicted on the "jacket." The *Pioneers* carry exterior plaques containing the same address, along with a depiction of two naked humans.

Voyager II will pass "only" 1.7 light-years from the small reddish star Ross 248 in 40,000 years. In longer time frames, the others will also pass various stars at similar distances. If very smart creatures from an unknown planet in those vicinities become aware of these dark, silent, schoolbus-sized objects (akin to our spying a boulder 44 million times farther away than the Moon) the schematic map could tell them where we live.

"Bad idea!" complained many, when launching the map was first suggested. Why give out this information to unknown aliens? Wouldn't it be safer to retain our unlisted number?

The Don't-tell-them-where-we-live controversy competed with another: the nakedness of the humans on the *Pioneer* plaques. The man was left alone, but the 5'6" woman was modified (over the objections of scientists) to look like an unclothed Barbie doll, and thereby less objectionable to those members of Congress who didn't like the idea of sending "smut" into space. If the aliens ever do discover what we *really* look like with our clothes off (How would they do *that*? Those medical abductions routinely recounted in tabloids?), they might want to understand that single alteration in an otherwise accurate diagram. Unintentionally, the omission will have communicated something significant about life on twentieth-century Earth.

We know one thing: Nature has created wonderfully clever ways of communicating, including numerous still-undeciphered methods employed by various species. From cave drawings to our 200,000-word English vocabulary, from the Rosetta Stone to the intricate equations of math and physics, we already have a full toolbox. What lies beyond would probably leave us speechless.

Chapter 11

Eat the Aliens

Speaking of aliens, only the gullible could think that an otherwordly invasion is in progress. Given an adequate science education and average skepticism, most citizens would adopt the mainstream view that extraterrestrials probably exist somewhere but are not actually seated among us at football games or darting among the clouds in flying saucers.

Scientifically, the widespread belief that extraterrestrials are out there stands on a tenuous base whose first serious treatment may have been the epic 1966 work of astronomers Carl Sagan and I. S. Shklovskii entitled *Intelligent Life in the Universe.* Sagan and Shklovskii employed an equation devised by their colleague Frank Drake of the National Radio Astronomy Observatory as a springboard for reasonably concluding that our galaxy contains a respectable number of advanced civilizations. The Drake equation multiplies the number of probable planets in the Milky Way (somewhere between 10 billion and 2 trillion) by the fraction that are Earthlike (possibly 3 percent) and a few other factors, such as the amount of time that an intelligent civilization might persist, finally arriving at a definite range of expected contemporary worlds inhabited by thoughtful organisms. Unfortunately, depending on which guesstimates are inserted into the equation's variables, that range can yield a total of *one* communicative civilization in the galaxy (ours) or 10 million. All in all, not too helpful at narrowing the possibility down, even if the process itself is fascinating.

Increasingly sophisticated astronomical techniques, along with bigger telescopes and faster computers, are expected to be able to identify Earth-like planets before the year 2012. We may then be able to define one of Drake's important unknown factors—the actual number of worlds that could support life. But even now there are tantalizing hints that life may be prevalent rather than rare. Amino acids, the building blocks of earthly organisms, have been identified on comets and found in meteorites, and one (glycine) has been spectrally detected in distant nebulae. Even so, we're left with very little more than speculation, and this brings out the armchair philosopher in all of us. For many, the belief in extraterrestrial life arises largely because the alternative seems implausible: that the universe's inventory of 200 billion known galaxies is just an immense backdrop for a single large-brained species whose leisure activities include stock-car racing and food fights.

So we search. Finding any sort of life would elevate exobiology to a science whose subject is real rather than imaginary. Lacking hard data, we meanwhile indulge in endless guesswork.

Much of our collective mindset comes from sci-fi novels and movies, which generally present aliens as being either benign and cuddly (like Spielberg's ET) or obsessed with destroying us. Delving into fiction may seem pointless, but since most working models for extraterrestrial visits have been powerfully molded by such exposure, dissecting those widespread paradigms may be useful, if only to discard the patently untenable ones.

So what would aliens really be like? In the sillier fiction, they're sometimes squidlike things that take over the bodies of everyone in the neighborhood, a process that understandably changes people's personalities, usually by removing their sense of humor. But entering human bodies seems an odd thing for visiting creatures to want to do. If *we* discovered slimy squids on some other world, it's hard to imagine that our first impulse would be "Let's crawl into these disgusting things and see if we can become squids, too." Sometimes, though, the aliens are gentle, and their desire to obliterate humans stems from some larger benevolent motive (for example, to stop our destructive activities like air polluting and futile chronic dieting). Or else they want to welcome us into some sort of benevolent interplanetary league.

While movie aliens are invariably either sweet or sinister, actual ETs might be neither. They could arrive out of simple curiosity. They could look around without bothering to let out a peep and then blast off into the distance. We'd be left thinking, What was *that* about?—and spend the rest of our lives squirming with anxiety.

Or they might be interstellar cult members who have come here to proselytize, in which case they would fly in, distribute a few billion leaflets, and then dash around recruiting followers. If they hovered without visible support or opened doors just by looking at them, it would certainly hold our attention long enough for them to sermonize. (If we could glimpse our own technology a mere century from now, it might seem just as indistinguishable from magic.) And no question about it, they would find human recruits no matter how loony the creed.

A real problem could arise if we found them irresistibly scrumptious—or vice versa. If aliens tasted like lobster, the SETI (Search for Extraterrestrial Intelligence) project, which has been monitoring stars for radio transmissions, might become a Search for Extra Tasty Individuals. Humans have long believed in "lunch first, science later."

Or the visit could be prompted by something totally unexpected, such as the aliens being intergalactic jokesters looking for new material. Even infants smile—perhaps humor is universal. They might bring us flying saucersful of whoopee cushions and chattering teeth—novelty gags that experts concede would have widespread appeal throughout the universe. They cruise the galaxy to crack up other civilizations, to boldly jest where no one has joked before.

Or maybe they just want to talk. Wouldn't *we* love to find some species that could serve the role of bartender while we blabbed on and on? Why shouldn't other life-forms also need to bend someone's ear? And what if they were a loquacious culture that simply wouldn't shut up? If they'd been drinking tons of their coffee-equivalent to stay awake during the long haul from Tau Ceti, we might not be able to keep up with them. As with any conversation, the small talk grows stale, the news gets old, and the aliens eventually depart for better gossip. But taciturn extraterrestrials would be even worse. Our unstated goal, continually reinforced by fiction, is to find life whose penchant for communication more or less matches our own—a tough assignment.

As for motives, what *really* sets Earth apart is our abundant water. We might well attract creatures who had no interest whatsoever in interplanetary relationships: They would view our continents as just the irrelevant stuff between the coveted oceans, our world a vast underdeveloped aquatic theme park.

All such speculation may seem pointless—and it is. There's not much we can honestly accomplish with just the knowledge that life dwells on a single world. Should an interplanetary exchange ever take place, it would probably resemble nothing in our wildest, too-much-pizza-before-bedtime dreams.

One indicator of how easy it is to jump the gun regarding imagined ETs is our history of false alarms. In 1967, signals that seemed artificial were detected by the British radioastronomer Jocelyn Bell at Cambridge University, using the university's state-of-the-art radio telescope, which she had helped to build. She and her mentor, Anthony Hewish, eventually decided that the LGMs (Little Green Men, which they'd first thought they'd found) were instead a new type of celestial body—a spinning neutron star, quickly dubbed a pulsar, which emits precisely timed flashes with each rotation. Precision and repetition are indeed a human sort of signal. But not in this case.

Then, in 1996, the scientific community was rocked, literally. Fossilized bacteria were reported imbedded in an Antarctic meteorite with the catchy name of ALH84001, which originally came from Mars and landed here 13,000 years ago, during the last ice age. Current majority opinion: Another false alarm.

Where might our first genuine extraterrestrial biota turn up? Estimates of the nearest likely aliens now place them in the seas beneath the floating ice sheets of Jupiter's moon Europa. If so, we'll have to go there to see them, since such creatures would be firmly aquatic and probably exhibit the same desire to hurl themselves through space that catfish do.

For all these planets and possibilities, public perception still associates extraterrestrial life only with Mars. We now know enough, however, to rule out advanced life there, unless it exists underground. Since subterranean creatures would surely find surface life dangerous and uncomfortable, we can assume that any Martian invaders would confine themselves to our subways and underground malls. (When you look around it sometimes seems as if they're already here.)

The nearest invaders may be *very* near: right here in our brains. Sometime in the distant past, strange microscopic creatures called mitochondria, which resemble no other earthly organism, succeeded in taking up residence inside our bodies. Nowadays we don't give them a thought: It's a useful symbiotic relationship. We provide them with living quarters within every cell, and they deliver a potent energy source called ATP. For all we know, mitochondria or their biological ancestors could have arrived here aboard long-ago comets.

Invaders that have become indistinguishable from ourselves? Why not? The origin of earthly life remains nearly as mysterious as that of any extraterrestrial organisms that might sail the still uncharted depths of the cosmic sea.

Chapter 12

Out of the Blue

If you stare at the blue sky for any length of time, you'll attract the wrong kind of attention. But do it anyway. Hard to believe perhaps, but unexpected treasures materialize where few are seeking them—out of the blue.

The visual phenomena to be explored in this chapter all owe their existence to a single source, the Sun, whose power output is astounding. The Sun expends 4 million tons of hydrogen per second in its ongoing conversion of mass to energy. This is no mere theoretical idea: If the Sun could be placed on a giant scale, it would weigh 4 million tons less every second. This prodigious production, this geyser of colors spewing from the solar core, has its strongest emission in the green part of the spectrum, which is why our eyes, in response, are most sensitive to green light. The sun's intense bombardment in all the visual wavelengths provides the direct or indirect source of every visible phenomenon in our lives.

Daytime sky observing requires neither equipment nor astronomical knowledge (no need to be familiar with constellations, for starters), and most of the diurnal phenomena are curiosities you've always seen but probably never recognized.

The Blues

Start with the blue sky itself, noting the varying shades in different directions. The darkest region is normally directly overhead, because that's where we're peering through the thinnest atmosphere. (You gaze through twice as much air when you look one-third of the way up the sky from the horizon.) But there is also a second dark band, 90° from the Sun, where light is maximally polarized—that is, where much of the light reaching our eyes is vibrating in a single plane. At dawn and again at dusk, those two dark areas coincide. The result of this union is a wide, deep-cobalt belt that crosses the sky from north to south at the zenith. This deep-blue band (which further intensifies through polarizing sunglasses) hosts the half-moon a few days each month—the phase offering the best lunar viewing. Through binoculars or small telescopes its shadow line (called the terminator) directly faces the observer, causing optimally shadowed mountains and craters to meet our eyes with minimum foreshortening. Craters are now not only perfectly illuminated but appear round instead of elliptical. (The two half-moons are officially called the "first quarter" and "last quarter," a source of understandable confusion, since only in astronomy can a quarter and a half mean the same thing.)

Another unexpected clear-day oddity: the sky's shape. It is, of course, a dome. Since you, the observer, are centered in that hemisphere, the section directly overhead should look as distant as the sky at the horizon. Does it? Check it out any clear day. The illusion is obvious: The overhead sky seems much closer to you than the horizon regions. A cloudless sky appears to be a flattened bowl and not a hemisphere at all. Try to point halfway up the sky. Nearly everyone gets it wrong and aims much lower than 45° from the horizon. The cause is our lifetime visual experience with clouds; overhead clouds *are* closer!

Daytime Stars

Stars in the blue sky? Yes and no. The starlike planets Venus and Jupiter are indeed visible by day (and Mars when it makes brief, rare, extremely close visits as in the summers of 2003, 2018, 2035, and 2050), but no true stars. The popular notion that stars materialize when you gaze up from the bottom of a chimney is a myth, probably circulated by chimney sweeps as an effective flue-cleaning reminder.

Jupiter is brightest around its yearly opposition, when it rises at sunset. However, daytime observers get a "second coming" three months later (in a configuration called quadrature), when it sits in that darkest zone of the daytime sky at right angles to the Sun.

Venus never appears in that dark day-sky district, since it can't wander more than 47° from the Sun. It compensates with unique brilliance. The cloud-covered world assumes its optimal appearance at nineteen-month intervals and retains its good looks for about three months. You can easily find it in the daytime with the naked eye by counting off four arm's-length clenched fists (each measures 10°) to the upper left of the late afternoon or setting Sun, and there it is! Stand where the Sun is blocked, as by a tree or the corner of a building; otherwise your pupil will constrict so much in the direct sunlight that you'll find it much more difficult to see Venus leap into view against the blue sky. It's worth the bother: People are always amazed to see a distinct "star" in the daytime sky.

When to see a Daytime Venus
(Upper left of the low afternoon Sun)

January–June 2004

September–December 2005

March–August 2007

December 2008–March 2009

Sun and Shadow

Now that solar storms have waned after their last maximum, in 2000–2001, opportunities for observing them are more precious; it's a good idea to use a pair of inexpensive welders' goggles. Be sure you get shade #12 or #14, for safety's sake. Through them, naked-eye sunspots pop into visibility if they're Earth-sized or larger. The Sun is always impressive when viewed this way.

Alternatively, and somewhat more crudely, you can utilize the pinhole-camera principle. The ground all around us contains countless overlapping images of the source illuminating it, which by day is of course the Sun. By twisting your fingers to create a tiny opening in your cupped hands, you can project the Sun's image onto any white surface.

Blue and Gold

The most amazing daytime fact is also the simplest, and it explains things that we should have learned as children but didn't. It's generally assumed that twilight colors are mysteriously produced by some sort of atmospheric legerdemain. Wrong. The colors do not originate here at all, but are the actual colored photons produced in the center of the Sun. Remember, the Sun emits no white light; it radiates the colors of the spectrum. Any scrambled mixture that contains red, blue, and green will be perceived as white. At sunset, the path that sunlight takes through the different thicknesses of our atmosphere bends, or refracts, the light, with each color at a distinct angle, allowing us to see the Sun's true colors side by side instead of mixed together.

There's more. We see a blue sky by day because the Sun's shorter (blue) wavelengths get scattered by air molecules six to ten times more readily than other colors. But that blue color has to come from somewhere: It is light that has been subtracted from sunlight en route to our eyes. In other words, the sunlight we see is always missing some of its blue, which has gone into manufacturing the blue sky. This explains why the Sun itself appears lemon-tinted. Although astronauts and high-altitude balloonists report that it's really a pure white star, the Sun seems yellowish to us here on Earth because its white light has been robbed of some of its blue. This intimate but seldom recognized tie-in between the yellowish Sun and the blue sky is both surprising and logical. Blue sky, yellow Sun. We wouldn't have one without the other.

Interestingly, the two colors can readily recombine. A snowy field on a clear day (or a white sheet spread on the ground) is illuminated by both the yellowish sun and the blue sky; therefore the reflection of the two, reaching our eyes together as a mixture, contains both elements. So the true color of the Sun, the color perceived by astronauts, is precisely recreated by sunlit snow.

Halos

Several strange but dramatically vivid atmospheric phenomena owe their origin to sunlight interacting with bits of water or ice. Probably the most common of these is the halo, a ring around the Sun or Moon that is always the same consistently huge 22° in radius, which neatly matches a maximally spread-apart thumb and pinky tip held at arm's length. Caused by sunlight or moonlight refracting and reflecting off flat hexagonal ice crystals, halos are so much larger than expected that they often escape notice (especially since few people look up toward the Sun in the first place). Halos around the nearly full moon are also common, but the solar version, exactly the same size, offers more vivid colors. Red is always on the inside—the reverse of the sequence seen in a rainbow—and the halo usually turns whitish outward of that. We rarely see the full spectral colors.

One or both sides of the halo often feature a dramatic bright spot that is sometimes white, sometimes vividly colored. These are parhelia, or sun dogs, which commonly appear by themselves with the halo absent. Halos materialize when high thin clouds overspread the heavens, with sun dogs additionally requiring that the Sun be rather low, an hour or two before sunset.

Over the Rainbow

Very nearly twice the size of a halo, a rainbow is an enormous arc 42° in radius. Its exact center is *always the shadow of your head.* During a sun-shower, or when rain falls in a place opposite the Sun from you, a rainbow *must* appear as long as the Sun is noticeably less than halfway up the sky. The sunlit water droplets can come from a sprinkler a few feet away or a thunderstorm in the next valley; the rainbow's apparent size remains the same regardless of the distance to the sunlit drops.

A rainbow cannot be seen if the Sun is higher than 42°, a rainbowless period that persists in most of the United States and Europe from mid-April through August between 10:00 A.M. and 4:00 P.M. The largest bows materialize when the Sun is lowest. At sunset, when the shadow of your head is projected to the horizon, the rainbow can be a full semicircle that spans almost half the distance to the zenith.

Sometimes a second rainbow appears above the first, always with dimmer colors whose order is reversed. This secondary arc lies 9° above the primary bow and is caused by a second reflection within each raindrop; the additional light-bounce accounts for both the dimness and the color reversal.

Occasionally you'll see a series of alternating bright and dim bands (usually aqua and pink) just beneath the main rainbow. These are called supernumerary arcs, caused by light-wave interference.

Would a rainbow exist if you were not present?

The answer is best visualized by picturing the rainbow's geometry as occurring on the surface of a cone whose apex is your eye and whose axis runs from the Sun through your head to your shadow. (As noted, it doesn't matter where on this cone the sunlit drops of water lie.) If you were absent—if there were no eye or camera at the apex—no rainbow would exist. Or, rather, each sunshower produces billions of potential rainbows, each slightly offset from the next. Complete the geometry by existing at any location, and you perceive one of these rainbows. A person standing next to you always sees an entirely separate rainbow, because its reflection and refraction come from a different set of droplets.

Here's yet another way of visualizing the situation. When you look at a specific spot on the rainbow—say, in the red zone—you're gazing at raindrops that are actually refracting all colors in slightly different directions, but you're in line to see just the red. Lower your head and you'll intercept the green, blue, or violet emitted from these same drops.

Other rainbow oddities:

The sky outside the rainbow is always darker than the sky inside. This vivid murky phenomenon is called Alexander's Dark Band—the largely forgotten inspiration for both an old Irving Berlin song and the name of a current rock group.

The reason a rainbow's ends terminate at the earth—why it's only a semicircle—is this: Rain tends to stop when it reaches the ground. But if you gaze downward toward a waterfall from a higher perch, you may well see a rainbow that's more than half a circle.

The Mystery of Color

The colors that paint a rainbow or twilight's clouds are photons that took a million years to squirm out of the solar core and escape the Sun's gassy surface, or photosphere. They then spent all of eight minutes of their newfound freedom whizzing through space to our eyes. We could call these the Sun's "real" colors: violet, blue, green, yellow, orange and red. All other hues, like purple, pink, aqua, and maroon, result either from mixtures of spectral hues or from light waves diffracting or interfering with each other (like the cloud iridescence noted in chapter 2). So anytime you see a non-spectral color, like the round brownish smear ("corona") encircling the sun or moon when behind a warm thin cloud, you can bet that some physical hanky-panky has transformed the light's journey.

In addition, the colors we perceive are intimately linked to the architecture of the human eye. Some combinations work for us, some don't. Mix red and green light and we perceive yellow. Try as we might, we cannot see "reddish green" or "bluish yellow." But our neural architecture does allow us to perceive reddish blue and bluish green. And other unknown and unperceived colors, with wavelengths shorter or longer than our retinas can sense, are out there, too. Insects perceive shorter wavelengths, while our skin "sees" longer wavelengths than our retinas do, experiencing these as the sensation of heat.

You're in a huge celestial art gallery whenever you leave your house, now that you know what to look for and why you're able to see it. The habit of daytime sky-gazing may be a bit eccentric, but it may help to banish the blues.

Chapter 13

Sky Spectacles

Our addiction to pleasure propels us to do risky things like marrying on impulse or sitting trackside at NASCAR races. We're fascinated by the unusual, and it needn't be weird or violent to qualify. Any sort of extreme will do, and when this "ultimate" happens visually in the natural world, it moves the human spirit in a way that is universal and unanimous.

To surpass the field of glorious contenders, to top the cosmic charts, a phenomenon must not simply be lovely, like Orion's return each autumn, but a step beyond, an Oz journey to the truly mind-blowing. And it must be rare. If our sunrises were like those on the Moon or Mercury, a nearly instantaneous change from gloom to brilliance, then an unexpectedly multicolored dawn would be wondrous beyond words. If such a marvel happened once per millennium we'd build cathedrals and legends around the experience; no one on Earth would miss a moment of it.

The near unanimous vote, the premier sky spectacle, has to be the total solar eclipse. It's caused by the astonishing coincidence that the Sun and the Moon—the only disks in Earth's sky—appear exactly the same size. This singular event shouldn't be confused with partial eclipses, which are viewed by millions of people every year. Totality is as different from a partial eclipse as night is from day—literally. As if to ensure that this greatest of spectacles remains the most unusual, there is on average only one total solar eclipse every 360 years for any given earthly site. If it's cloudy, you have to wait another 360 years. Unfortunately, the United States is suffering an unprecedented thirty-eight-year eclipse drought. There hasn't been a totality anywhere in the lower forty-eight since 1979, and there won't be another until August 21, 2017. But totality occurs somewhere in the world nearly once a year. You'll have to travel.

Worth the journey to Mongolia or Tierra del Fuego? The emotional reactions of all who attend is answer enough. They groan, moan, shout, and exhibit behavior generally seen only at Christmas office parties. It's not just the dogs that bark when the Sun, Moon, and Earth form a perfectly straight line in space. You *sense* something happening, to a degree that a blind person might well be aware of. Totality seems to send one's nervous system all the way back to Cro-Magnon.

The Sun's glowing corona presents a unique, ultrahot, million-plus-degree radiance that dramatically leaps across the sky on this, and only this, occasion. Because its inner segments are bright and its outer regions faint and lacy, photography fails to capture the entire coronal apparition. Yet the human eye, with its sensitivity to a wide brightness range, manages it with ease. It is for this reason that no photograph or televised image can remotely capture the look and feel of a total solar eclipse.

As if this weren't enough, various ancillary phenomena materialize. "Shadow bands" are especially noteworthy. Shimmering streams of curving black lines ripple across any white surfaces on the ground, such as snow or sand, some sixty seconds before and after totality. Incredibly, they cannot be photographed! Arresting to the naked eye, they're simply not present when you later look at the developed snapshot or play back the video. Shadow bands have such low contrast that they do not reproduce photographically, whereas the human eye can construct a subtle image if the phenomenon is moving. Even if you didn't know that this dramatic spectacle is not a Kodak moment, just looking at shadow bands raises goosebumps.

The minutes leading up to totality also create bizarre lighting on all earthly objects. Because sunlight now emanates only from a starlike spot on the Sun's limb, normal shadows are replaced by ultrasharp ones, with colors assuming otherworldly tones and strange degrees of saturation and contrast. In short, *everything* looks strange. And then there's the eclipsed Sun itself, safely viewed only with eye protection (such as welders' goggles shade #12 or #14) until totality begins, at which time it can be observed directly with the naked eye.

People have been known to take second jobs or mortgages, do whatever they must, to once again place themselves in the shadow of the Moon. Since totality may last anywhere from one second to seven minutes and travel expenses are usually high, the per-minute cost often runs to four figures. Worth it? Here's how to decide for yourself—with the dates and rough locations for the next ten totalities. Also included are an appraisal of viewing prospects, based on the climatological odds of clear skies and the likelihood that, if it's an oceanic eclipse, cruise ships heading there will not be going so far from their normal routes that they must charge astronomical fares:

April 8, 2005, South Atlantic—poor
March 29, 2006, Libya, Eastern Mediterranean, Turkey—excellent
August 1, 2008, North Pole, Siberia, China (Great Wall)—fair
July 22, 2009, India, China, Western Pacific—excellent
July 11, 2010, South Pacific—fair
November 13, 2012, South Pacific—good
November 3, 2013, South Atlantic off Africa—poor
March 20, 2015, Arctic Ocean—poor
March 9, 2016, Indonesia, Pacific Ocean—fair
August 21, 2017, United States, coast-to-coast—excellent

Amazing Sky Spectacle Number Two is even more infrequent—the meteor storm. Picture sixty shooting stars radiating each second from one point in the heavens and continuing for more than an hour. Such displays are like a wildly sputtering short circuit from another dimension, and they generate as much bewilderment and fear as wonder and joy.

Several years ago, one of my students told of having unexpectedly observed the meteor storm of November 17, 1966. She was in a train crossing the Texas countryside, and she was sure that the end of the world had arrived. What else could this be? Looking around the compartment, she saw that everyone else was sleeping, unaware that the heavens were exploding. She anguished over what to do: Should she awaken these strangers? What was the correct protocol? Do you rouse people to witness the end of the world or let them die unawares? Then a conductor came by and together they watched the astounding phenomenon until dawn brightened the sky.

After the 1966 storm, the next such excessive display was due around the turn of the century, and it arrived in an unexpected way. November 1999 did see a brief but marginal storm in Middle Eastern longitudes. The following November's "storm" was even skimpier. Then came the November 18, 2001, thriller. People throughout North America and extreme eastern Asia gaped at brilliant meteors streaking across the sky at the rate of fifteen per minute, nearly all trailing lingering greenish streaks. Though the display was just a pale approximation of the one in 1966, early risers nonetheless witnessed the finest meteor display of their lives. *That* qualified! A year later, on November 19, 2002, a final predicted storm was generally perceived to be a bust. Next probable date: November 2099.

After eclipses and meteor storms, the prize has to go to a Great Comet. Two crossed our skies in consecutive years, 1996 and 1997, when Comets Hyakutake and Hale-Bopp graced the heavens—both, oddly enough, in the month of March, just like the previous noteworthy comets, so-so Halley in 1986 and beautiful Comet West in 1976. In dark skies, the fainter Hyakutake was the more spectacular of the two with its sky-spanning presence, although far more people witnessed short-tailed Hale-Bopp because of its greater brightness. Neither was remotely as dramatic as some of the mind-numbing comets of history, such as the Great January Comet of 1910.

A "Great" comet is not usually discovered until just a couple of months before it becomes spectacular. Then it stretches motionless across the heavens, sometimes spanning half the sky. Illuminated by sunlight but also glowing on its own, its presence is so strangely compelling that you immediately understand why cultures throughout history feared and revered such objects and inscribed their likenesses on cave walls and tapestries. Nobody can predict when the next new one will arrive, though they average about fifteen years apart. The only short-period or predictable comet that can put on a supershow is the famous comet Halley. Unlike its disappointing visit of 1975–76, the geometry is right for a true spectacle when it next arrives in 2061. Its best display in 2,000 years should unfold on the following visit, in 2137.

Everyone who has seen a bright, sky-filling display of the northern lights would place it on the "best spectacles" list or even award it top position. The aurora borealis is commonly associated with penguins or polar bears, but the waving curtains of greens and occasional reds are not confined to distant frozen landscapes. The eerie spectacle often appears over dark skies throughout the northern half of the United States and sometimes even spreads its delicate green fingers over Mexico!

While they are dramatic and often spooky, the northern lights usually display little or no color, most commonly a pale green (at 5,577 angstroms wavelength), caused by excited atoms of oxygen a hundred miles up, or more rarely a deep red generated by the same process. (Interestingly, a red aurora is precisely the same color as a laser pointer; both are 6,300 angstroms.)

In upstate New York, where my observatory is located, displays of the northern lights sometimes shimmer crazily across the entire heavens, producing new configurations each fraction of a second. The extraordinary spectacle of April 6, 2000, was brilliant enough to appear through the lights of New York City and enveloped the Moon in rich red blobs of dripping paint. It materialized everywhere *except* in the northern sky! Such mind-benders happen a couple of times a decade. While they occur far more often during times of peak sunspot activity (next expected from 2008 to 2012), they can and do erupt anytime.

Alchemists of surprise, auroras can exhibit blotches, rays, bands, arcs, curtains, or any combination. They can appear to float supernaturally motionless, or pulsate leisurely, or flicker rapidly enough to produce five patterns per second, the observer's spirit seeming to vibrate along as if in chorus. The profound silence accompanying such a frenzied visual eruption is eerie. And so are occasional reports of crackles or hisses, attributed to the disputed human ability to sense the huge electrical charges on the ground beneath them. (During three winters in central Alaska in which I led a series of aurora-observing expeditions, numerous natives told me of definite noises accompanying the lights. These people had not the slightest doubt, although University of Alaska researchers have never detected them despite their many attempts using sensitive microphones.)

Residents of the United States are lucky. Northern lights are sections of an extensive glowing halo centered over Earth's magnetic poles (to which compass needles point, as opposed to the geographic poles around which the planet rotates). Through an accident of geography, the magnetic pole lies exactly north of the central United States, some 13° of latitude closer to us than the pole of rotation. People at a comparable latitude in Europe or

Asia live 1,800 miles farther from this auroral oval and experience far fewer displays. Rome is the same distance from the geographic pole as New York but sees auroras as infrequently as Miami.

We now fully understand an aurora's genesis. It starts with a violent mass ejection from the Sun's corona—a blast of electrically charged particles that sweep across Earth's magnetic field and influence it if the material's magnetic polarity happens to have the opposite orientation as Earth's. This blast generates electricity that excites electrons in the oxygen atoms of our upper atmosphere. Serious power: 20 million amps at around 50,000 volts are typically funneled into the regions above the magnetic poles, setting their atoms aglow like fluorescent tubes. If Earth had circuit breakers, they'd all trip.

And in a way, they do. As if to echo the sky action, electrical currents surge through the ground, especially along pipelines. They whip through transmission and power grids. On March 13, 1989, during an intense aurora seen all the way to Mexico, such sudden currents destroyed giant transformers; nearly the entire province of Quebec was plunged into night-long darkness while garage doors throughout North America kept opening and closing from midnight till dawn.

Every month or so a faint aurora is visible from the northern United States, and once every year or two a fabulous whole-sky event blazes across the heavens. Great auroras can and do happen anytime in the solar cycle. If you live in a rural area away from light pollution and occasionally glance northward, even while you're walking from the car to the front door, there will come a night that will present you with the treasure of the fabled lights.

The next finalist on the "All-Time Greatest" list is the most ephemeral—a bolide, or fireball. This is an exploding meteor, a large fragment of icy rock or stone of sufficient speed and mass (less than an ounce is enough) to either break apart serially like a Roman candle, or explode outright from ultrahot interior gases. Often it possesses enough brilliance to cast shadows or rival the full moon. No one can predict when one will rip through our atmosphere, lighting up the countryside and startling onlookers a few times a year. Unfortunately, except for insomniacs, truckers, rock musicians, and workers on the graveyard shift, few people are usually awake to see them. As a colorful bolide overwhelms the heavens (emerald green is a favored hue), trailed by fiery cascading fragments, it may commonly manifest as a sudden glow behind bedroom curtains, shrugged off as "heat lightning." Those who have seen one never forget it—and marvel that so few other people witnessed the wonder that passed in the night.

The final memorable sky apparition could easily be a double halo around the Sun. Meteorological rather than astronomical, it's a spellbinding sight that an avid skywatcher might see only once in a lifetime. When it occurs, the Sun is surrounded by two brightly hued concentric circles. The inner ring is always the common 22° halo seen among cirrus clouds a few times a month. The outer one, much rarer, boasts an astounding 46° diameter slightly more than twice the inner ring's radius and potentially stretching from horizon to zenith!

If you never see a double halo, let's offer a second, more common ice-crystal wonder as an alternative, one that appears several times each year: the circumzenithal arc. It's vivid and usually puzzling, with its ends pointed upward and its bow aimed downward at the Sun. Materializing only when the Sun is less than a third of the way up the sky, the circumzenithal arc looks like something that ought to be impossible: an upside-down rainbow suspended in the air. Unlike halos, this amazing apparition displays the full intensity, saturation, and complement of rainbow colors, but with the arc centered directly overhead (instead of centered on the shadow of your head, like common rainbows). Moreover, it's often seen against a blue sky. No sunshower required, just the thinnest veil of lacy ice-crystal clouds.

Telescopes can expand the list of great sky apparitions. Nobody looking through a good instrument at the half-moon, Saturn, or a globular cluster (arguably the three most viscerally powerful telescopic sights) can fail to be amazed. But since these half-dozen "best-of-the-best" materialize to the naked eye without technological intercession, they convey an especially powerful impact.

These various wonders may linger for weeks or vanish in a few seconds. Some occur without warning, a gift reserved only for those who care enough to watch the sky. Dedicated professional astronomers, who spend most of their careers indoors at computer screens, are not among the lucky ones who generally witness such spectacles. People whose jobs keep them outdoors have the edge. The rest of us can join this high-probability pool merely by habitually elevating our noses.

Chapter 14

The Shadow Knows

The strangest of our national traditions? Hint: It also enjoys the longest history and weightiest celestial implications. Hard to believe, but it's February 2. Groundhog Day.

Extraterrestrials arriving some February to study our customs might find Presidents Day helpful in understanding the "Faces on Earth" they've viewed telescopically on that peculiar midcontinental cliff face, which their experts insist are natural formations. But Groundhog Day?

Let's explain it to them simply. Woodchucks emerge from their burrows and then, despite having IQs of 0.7, look for their shadows cast by the rising sun. If they do see their shadows, they suffer panic attacks and dive back in. That sequence of events triggers Earth's Northern Hemisphere to produce six more weeks of winter. And this entire scenario revolves around the perceptions of a single Pennsylvania groundhog, who serves as the entire country's spokesrodent.

Beyond the underfunded science of woodchuck anxieties, there's the suspicion that the whole thing is rigged. Six more weeks of winter? Surprise: There's always that much more winter. Since the vernal equinox forever falls six and a half weeks after Groundhog Day, spring's onset reliably occurs that much later whether or not rodents are on mood-controlling meds.

And yet Groundhog Day is neither arbitrary nor unrelated to the cosmos. It actually connects with astronomy. February 2 is originally Candlemas, the fortieth day of Christmas, one of the year's four "cross-quarter" days, which are exactly centered in the seasons. Groundhog Day is midway between the winter solstice, when the Sun rises at its rightmost extreme, and the vernal equinox, when it ascends due east. Marking the halfway point of those two yearly milestones, a cross-quarter day provides the calendar with a badly needed tick mark, just as "southeast" helps out a compass by filling in the place between east and south. Overlooking the strange recent introduction of marmots into the custom (which itself comes from an old German superstition), that cross-quarter day of February 2 was an important anniversary for centuries on end.

Traditionally, that is. The precise seasonal midpoint has now shifted a couple of days, to February 4, because of the wobble in Earth's tilt known as the precession of the poles, which swings Earth's axis around in a complete circle every 25,780 years. The tweaking of the calendar brought about by the Gregorian reform in the sixteenth century, correcting an accumulated discrepancy, further created new dates for the solstices, equinoxes, and their midpoints.

Besides the "six more weeks of winter" business being rigged, the groundhog challenge is also bogus because February is one of the cloudiest months in the Northern Hemisphere, according to long-term averages. As astronomers often lament, the odds of seeing the sky are then lowest, along with the chance of a groundhog seeing his shadow. Moreover, the Sun is far less likely to be visible at around sunrise (or sunset, for that matter), because clouds near the horizon are viewed horizontally and tend to "fill in" intervening clear spaces. Obviously the deck is fully stacked against any shadow viewing.

So the statistically likely outcome is for Punxsutawney Phil to see nothing and for winter to end before six more weeks are out. Alas, this is just an empty promise for residents of Pennsylvania, where the farce is held, and for most of the United States and Canada. Most inhabitants of North America would gladly accept a mere six more weeks of winter. For them the soonest realistic early spring will occur only when Earth's axis swings halfway around—sometime near the year 13000.

Much intriguing information lurks behind shadows, well worth exploring any day of the year even if you're not a rodent. Artists correctly perceive that on a clear day, shadows are blue, not black or gray. That azure hue is obvious in those tree shadows on the snow. Why? Not much mystery here. Shadows are sunless zones whose only illumination comes from the blue sky.

When a tree's shadow is cast on a light surface, like sidewalk or snow, the shadow of the lower trunk is well defined with sharp edges, but the upper branches and twigs are blurry. Why? The underlying physics is straightforward: Shadow qualities depend on the size of the light source and the distance to the surface on which it's cast. Moreover, light waves bend around objects in a process called diffraction. (The combination of these two factors will cause a high-flying jet's shadow to vanish before it reaches the ground!) More shadowplay. As noted in chapter 13, the ground is covered with smeared, overlapping images of the light source, the Sun, and certain techniques can dramatically resuscitate the solar picture. Scrunch your fingers to make a tiny opening, a pinhole camera, and project the Sun's disk a few feet downward onto the ground. Position this Sun image so that it lies among blurry shadows of distant trees, and it will become a window that contains razor-sharp photographs of the faraway branches, projected upside down as if by an actual lens. Experiment with your pinhole camera while walking, for amazing results. But be prepared to do some explaining if neighbors are watching.

During partial solar eclipses, the Sun's crescent shape is easily projected in this manner. The myriad small openings between tree leaves always create hundreds of crescents on the ground during eclipses, nature's own mad special-effects department. San Francisco artist and Exploratorium demonstrator Bob Miller has established an entire artistic enterprise from this shadow business. He was the first to show audiences during science lectures that a pencil eraser hanging from a thread will cast a "negative" (black) image of the Sun on a sunlit sidewalk by essentially blocking out exactly one solar image from those scrambled together. During a partial eclipse, the shadow of the eraser is a precisely defined crescent—clearly not the shadow of the eraser but the negative image of the Sun itself. Eerie.

After the Sun sets, out pop the only two other celestial bodies in all the universe that can create shadows on Earth: the Moon, of course, and the third, least-known shadow maker, Venus. Nothing else, except for the kind of extraordinary supernova not seen since 1604, reaches the magnitude-4 threshold capable of casting shadows. Venus does twice, during each of its nineteen-month cycles from evening to morning star and back again. The brilliance lasts a few months and its shadows are unique. See the phenomenon for yourself on a dark rural moonless night by looking behind you at a white surface (snow, or a sheet spread out on the ground) and you'll be amazed at how sharp the shadows cast by Venus appear. Shadows cast by the Moon, for their part, look unexceptional; their ho-hum familiarity arises because they display exactly the same degree of medium-sharp definition as the Sun's, since both disks appear the same size.

Indoors, the blurriest shadows are cast by fluorescent tubes, while the sharpest come from ordinary clear (unfrosted) lightbulbs. Add color to the fun by setting up a triad of red, blue, and green floodlights (light's primary colors) spaced a few feet apart. The various shadows cast in front of this arrangement are sensationally vivid and colorful, and precisely create the primary colors of paint: yellow, cyan, and magenta.

Earth's Shadow

Now for the grandest shadow of all—the one cast by our planet. Most people will say they've never beheld Earth's shadow. In fact, they have. Often. They just didn't know it.

Every evening at dusk, a sharply defined blue-gray horizontal band overlies the eastern horizon and persists for fifteen minutes after sunset. It's not subtle and it isn't rare. The phenomenon lurks there every clear evening, period! Wonderfully named the twilight wedge, this is nothing less than our planet's shadow cast into space.

Above the upper margin of this band, rays of sunlight still streak out into infinity. Below, they're blocked by the body of Earth. If you are watching the sunset from the summit of a tall mountain (like Hawaii's 10,000-foot Haleakala and countless others), the mountain itself stands clearly visible within the twilight wedge as the peak's shadow is thrown off into space.

Earth casts its dark umbral shadow a million miles; happily, a single celestial body floats nearer than that and thus can receive the shadow to show off its full characteristics. The occasion, of course, is the lunar eclipse.

The Moon is the only known body whose speed matches its diameter: It alone moves through space its own width each hour. Thus it takes that long to fully enter Earth's shadow. During this time the Moon is said to go through a "series of phases," but it's really much stranger than that. Earth's shadow tapers like a chopstick to a little more than half its original width at the Moon's distance—still about twice the Moon's size. So a lunar eclipse is a geometric wonderland, involving curves with two different radii. This interaction of arc (Earth's shadow) and circle (Moon) produces its weirdest effects when the shadow falls near the lunar limb, or edge, around the beginning and end of the partial stages. To impress children and other impatient observers, have them look up then, during these Times of Maximum Strangeness.

Earth's shadow is not exclusively black. Our planet throws a colored shadow into space, since Earth's sunrises and sunsets combine into a brilliant ring that's bent, or refracted, into the shadow and paints the eclipsed moon. The color varies enormously. A totally eclipsed moon can be black, brown, gray, red, coppery, orange, or yellow-white, depending on the atmospheric dust, clouds, moisture, and pollution visible at Earth's limb, or edge. The lunar eclipse is thus the only celestial event in which we look at another world but get an environmental report card about our own.

Long before Columbus, the ancient Greeks correctly figured that our world must be a sphere, since that's the only shape that always throws a round shadow. They also realized that a lunar eclipse happened when the full moon went into our shadow, because eclipses occurred only when the Moon reached a point in the sky precisely opposite the Sun, which is where our shadow must lie! What else could an eclipse be?

The following is a list of the total lunar eclipses visible from most of North America from 2004 to 2019. If you know any Flat Earth Society members, here's when you can let them see our planet's shape for themselves:

2004:   October 27
2007:   March 3 and August 28
2008:   February 20
2010:   December 21
2014:   April 15 and October 8
2015:   September 27
2019:   January 20

Being the recipient of another celestial body's shadow is an even more powerful experience, and once more the Moon is the only such object near enough to fully oblige. (See dates of solar eclipses in chapter 14.)

Could it be that peering behind the science of shadows, this intriguing link between Earth and sky, is the hidden secret of Groundhog Day? Maybe they're onto something, those woodchucks.

Chapter 15

Egg-quinox

Throughout time, in cultures spanning the globe, equinoxes were celebrated with fanfare. Some civilizations threw wild parties, while more exuberant tropical communities tossed virgins or in-laws from monuments. Our own equinoctial announcements, broadcast on the six-o'clock news, barely cause a stir. Most people greet this sole vestigial trace of ancient astronomical practice with apathy or misconception: "Hey, they say eggs will balance on end today. Let's go to the 7-Eleven and buy some!" The media does its part by routinely repeating such myths as "Day and night are now equal." A few folks glance at the newspaper's local sunrise and sunset listings and see that day is longer than night at each equinox. They probably just shrug off the contradiction. After all, nothing in the papers makes sense anyway.

Here, then, is all you need to know about equinoxes, those last official bridges between human life and the sky.

If "March 21" pops to mind, you're dating yourself. You're probably over thirty-five. For North Americans, the final spring equinox on that date happened in 1979. Nowadays it's exclusively on the 20th. Reason: equinoxes slowly migrate earlier and earlier as each century progresses. Skipping the leap year at each century change normally corrects this slide. But a rare once-every-400-year calendar tweak befell 2000; that *was* a leap year, which let the equinoxes slip unchecked. Not to worry. It's all part of the straightforward sixteenth-century Gregorian calendar reform that only eleven people in the world understand.

At the moment of spring and fall equinox, Earth is angled exactly sidewise to the Sun. The Sun then hovers directly over the equator, above the Colombian town of San Miguel de Guacamole. (Actually I made that up. The spot on the equator lying precisely below the Sun at that moment changes from year to year.) The point is, neither pole is tipped toward that favorite star of ours, and, as the media never tire of reminding us, days and nights should therefore be equal.

But they're not. They would be if our planet were like the Moon or Mercury and had no atmosphere. But since the air between you and the horizon bends the Sun's image upward by its own diameter, making it rise a few minutes earlier and set that much later than it would otherwise, those extra five or so minutes of undeserved daily sunshine displace the true date of equality by nearly a week.

The equinoxes are a symbolic time of balance. From Penguinville to Irkutsk, the Sun rises and sets precisely due east and west. You can calibrate your sundial, that chore you've been putting off for so long. You can orient yourself and figure out which windows of your home face where. (The term "orient yourself" originally arose because if you knew which way the Orient lay, you could correctly point east.)

Here's a little-known phenomenon that even most astronomers seem unaware of: Only on an equinox does the Sun move in a perfectly straight line across the sky. A time exposure on any other day reveals the Sun on a curving path that is concave toward the north during the warm half of the year and toward the south during the cold months. But on March 20 and September 23 the solar track is laser-straight.

Can you figure out why? Outdoor enthusiasts realize that northerly or southerly stars, like Capella or Antares, forever arc in curving paths as night progresses, while those on the celestial equator, like Orion's belt, march arrow-straight across the heavens. Well, on the equinoxes the Sun sits on the celestial equator and thus follows the custom "When in Rome . . ."

The symmetry metaphor applies to so many facets of the equinox that it's no wonder some folks believe that eggs (symbols of life) will then balance on end. But the laws of gravity stand unrepealed on March 20 and September 23: Eggs remain as generally unbalanced as the obsessive egg positioners themselves.

It would be nice if Sun and stars could both be seen together at the spring equinox. (An equinoctial total eclipse might do the job, and we'll get one in 2015; unfortunately it's at the North Pole.) Then the Sun would clearly appear in front of a constellation that since ancient times has held special significance. People have always used that spot where the Sun hovers at the vernal equinox as the zero point for celestial longitude, the basis for navigation.

Trouble is, that point is slipping among the constellations. The 25,780-year precessional wobble in our axis is the culprit, causing the place the Sun occupies at the equinox to creep westward among the stars. Two millennia ago it was in the constellation Aries, which is why the vernal equinoctial spot is sometimes still called "the first point of Aries." In the year 69 B.C., however, it moved into Pisces, where it's been ever since.

If you've ever seen the rock musical *Hair,* you'll want to know when the vernal equinox will start happening in the constellation of Aquarius, to commence the Age of Aquarius, that golden period when supermarkets will again stock tasty ripe tomatoes. We shouldn't hold our breath. At the present rate of precessional advance (50.2787 seconds of arc annually) and taking into account its variable speed, the spring equinox will start to occur in Aquarius in the year 2597. Almost six centuries to wait before the age of harmony and understanding.

Of course, those who take the Aquarian Age seriously may object to these calculations. Constellation boundaries were not handed to us on stone tablets. Both Pisces and Aquarius are dim smatterings of stars that do not resemble, except to registered lunatics, the figures they're supposed to portray. The exact place where one constellation ends and another begins is arbitrary. The cited cross-over date is based on the internationally recognized constellation boundaries assigned by the International Astronomical Union in 1930: It does not belong to the natural world.

The equinoxes are in all other respects a milestone of nature, a time when Earth and sky are almost poetically united in an ephemeral moment of symmetry and balance. Worth celebrating. Maybe an omelet.

Tropic of Bull

The other major sky event is the solstice.

There are two, of course, but since the December 21 winter version often coincides with cabin fever and howling snow, let's focus on the pleasanter summer version, usually June 21.

Like equinoxes, most civilizations regarded the solstice as a big deal. Were they merely desperate for things to celebrate?

First the obvious: It's when sunrise happens at its leftmost spot on the horizon and sets at its farthest right. Early and late sunlight spills through windows at angles unseen at any other time, striking dusty corners of rooms that normally don't get illuminated enough to merit cleaning. The summer solstice also presents the highest possible Sun. Objects cast their shortest possible shadows. From nowhere in the continental United States, Canada, or Europe, however, is the Sun *ever* straight up. In mid-northern locations like New York, Denver, or Rome you will find the solstitial Sun 72° high at midday; it misses the zenith by 18°, roughly a spread-open hand at arm's length. The only place in continental America where the Sun blazes virtually straight up is Key West, where on June 21 it is offset from the zenith by a single degree (the size of a penny at arm's length). Essentially no shadows. Stark lighting.

The summer solstice also offers:

The year's longest day, with fifteen and a quarter hours of direct sunshine in a typical location—more, the farther north you go.

The year's longest twilight. Taking twilight into account, typical locales experience seventeen hours of bright usable daylight and just seven hours of night. Good for outdoor sports, bad for insomniacs.

Since the strength of the Sun depends on its height, the intensity of its visible rays and the power of its ultraviolet are maxed. Even blue and violet light are now intense enough to deliver burns.

But though the Sun is strongest because it's highest up, it's also unusually small. The Sun looks tiny because it's much farther away than average. Summer solstice happens just ten days from perihelion, the annual far point in Earth's orbit, when the Sun is 2.5 million miles farther from us than it is in January. This makes it look 3 percent smaller and causes its rays to be 7 percent weaker. In another 12,000 years, however, it will annually hover closest to us just when it's highest in the sky, producing horribly fierce summers and necessitating bigger air-conditioners in all new model 14004 SUVs. We're lucky that during our lease period the summer solstice happens when it does. Its timing is perfect for our comfort, with both warmth and sea levels near historic highs—just a few degrees and a few feet below their maxima 6,000 years ago. In fact, the changing dates of the solstice are intimately linked with intervals of global glaciation.

Climate variations are more complicated than the simple 26,000-year axial wobble, because Earth's orbit also becomes less elliptical in 100,000-

and 400,000-year cycles, while our degree of tilt—critical for producing and intensifying seasons—varies from 22.5° to 25° in its own 41,000-year periodicity. Meanwhile Earth itself reacts to differing degrees of solar insolation in various ways, while plate tectonics influences climate by creating such mischief as mountain chains that alter global wind patterns. Our global climate is nowhere near as clockwork-uniform as the simple precessional cycle.

With the Northern Hemisphere tilted toward the Sun, regions just above the atmosphere remain in sunlight throughout the night. The consequence is that orbiting Earth satellites, the space shuttle, and the International Space Station are easily seen for most of the night because they're gliding in brilliant sunshine against a dark, starry backdrop. You'll notice drifting starlike dots crossing the night sky every minute or two within eight or nine weeks of the summer solstice—between April 20 and August 20. This army of orbiters—mostly spy satellites that watch you even when you're ignoring them—are a reliable solstitial feature. You'll see them speeding every which way except toward the west. (It would be loony for NASA to launch against the direction of Earth's spin since it would require far more energy to attain orbital speed.)

In regions north and south of your home, the solstice is either more powerful or more inconsequential. In tropical countries the solstice is hardly noticed because it's almost meaningless. Tropical days and nights stay about the same throughout the year, and twilight there is always brief. Along with swaying palms and Piña Coladas, the tropical image is of long romantic twilights. Good public relations, but actually the Sun drops fast. Civil twilight (whose conclusion is defined as the Sun reaching a point 6° below the horizon, marking the moment for streetlights to come on according to most civil ordinances) lasts an hour in June in Montreal, but lingers only half that long in the Virgin Islands. Going the other way, farther north, in places like Scotland and Sweden and cities like Moscow, Copenhagen, Amsterdam, and Fairbanks, there's simply no night at all each June. Even at midnight the sky is still bright with twilight. Teenage couples looking for a dark place to park have to drive around until September.

The solstice has different connotations in various parts of the world. In most of south Asia, for example, June marks the *end* of summer heat, while south of the equator the June solstice denotes the start of winter. Moreover, many parts of the world have not four seasons but three: cool, hot, and rainy; solstices and equinoxes just don't mesh with their all-important climate scheme. But solstitial legends and myths abound in such places anyway. In India, for example, they believe that the most propitious time for spiritual progress is between the winter and summer solstices, not the other way around. Start meditating on June 22 and you've missed the boat. Start on December 22 and you're *Om*-free.

Another little-realized oddity: The year itself does not have two halves! You'd think that the period from equinox to equinox, March 20 to September 23, would be the same length as the other half. Pleasant surprise: Not so. Since Earth moves fastest when closest to the Sun—on January 3 or 4 (it varies a bit from year to year)—it experiences fewer rotations during that half of its orbit. So those of us in the Northern Hemisphere have five fewer days of cold. The calendar disguises this discrepancy by putting most of the short months, including February, into this "half" and also skewing the dates: The fall equinox happens later in its month, on September 23, but spring's lands on the 20th. Nobody in the Northern Hemisphere has ever complained.

With little hoopla, the solstice has just started occurring in the constellation of Taurus instead of Gemini, where it's been for the past 2,000 years. Beginning in 1992 and continuing for the next three millennia, the solstitial Sun will remain in the bull's arena, thanks to our planet's leisurely precessional wobble. But astrologers, who ignore precession and the actual constellations, say that the solstice is still in Cancer, just as it was in Ptolemy's day. Such a living-in-the-past attitude explains why the region where the solstitial Sun hovers directly overhead is still called the Tropic of Cancer. By rights, the signposts down there should be crossed out and changed to the Tropic of Taurus. This hardly affects the astronomical community, who regard astrology as superstition to begin with. But it may interest those with birthdays at that time. It can be puzzling to learn that your Sun sign is Cancer but the Sun is actually in Taurus.

No more so, perhaps, than to be told, "Welcome to the start of summer; now, today, days start getting shorter!" For greater solace, choose the December 22 solstice: At the moment winter arrives, days begin lengthening and solar intensity starts growing. Too bad it takes air temperature a month to catch up.

A bit of trivia: Since solstices will migrate earlier and earlier until February 29, 2100, the final December 22 solstice for North Americans will happen in 2011. After that it will be the 21st for the rest of our lives.

Doesn't matter much to us, maybe. But it did to the ancient Mayans, who would not have wanted to be sacrificed on the wrong day. A little screw-up like that and there goes your whole holiday.

Chapter 16

Deadly Light

After a day outdoors, sunbathers and skiers prize their "glow of good health," the rosy look painted by powerful ultraviolet energy. Since our lives can depend on reliable information about this invisible radiation, it's critically important to dispel UV myths and misconceptions. Here are dozens of vital facts and obscure curiosities about this dangerous lifetime companion.

The bad stuff begins just beyond the violet end of the spectrum. It may share the same name, but not all ultraviolet is equally dangerous; slightly shorter UV wavelengths crank up the peril big-time. A small difference in frequency produces vastly different effects: UV of 3,000 angstrom wavelength creates burns eighty times faster than UV of 3,200 A. (An angstrom is a hundred-millionth of a centimeter.) But even the longest and weakest—UV-A, from 3,200 up to 4,000 angstroms, some of which is actually visible to the eye as violet—has been decisively linked with malignant melanoma. Worse news: This longer-wavelength UV easily penetrates the atmosphere.

Public enemy number one, however, is UV-B (from 2,800 to 3,200), the main culprit responsible for sunburn and various skin cancers. This notorious UV is the only segment filtered by Earth's ozone layer. Even more powerful and lethal is UV-C, comprising everything from 2,800 down to 400 A, but this deadly superenergy doesn't penetrate the atmosphere at all: At a typical UV-C wavelength, only one photon makes it to the ground every 30 million years. Airless celestial bodies like Mercury and the Moon are continuously bathed in UV-C, which quickly breaks down chemical bonds and knocks electrons out of their orbits. Nothing, no unsheltered bacterium or bit of moss, could survive that sterilizing radiation for more than a few minutes.

Longer wavelengths of the electromagnetic spectrum—visible light, infrared, radio, and microwaves—can cause entire atoms to bounce around but are too weak to meddle with atomic structure. So cell phones and leaking microwaves, favorite media fear stories, can heat tissue but not induce cancer. Not that standing in front of a microwave tower is a good thing to do. Being exempt from a future cancer hardly matters if your brain is boiling. Microwaves make entire atoms bounce around, which is another way of describing heat. Every atom in our bodies constantly jiggles at 1,000 miles per hour, thanks to our 98.6° body temperature. (The idea of microwave cuisine was first grasped by military personnel who first found chocolate bars melting when they passed near radar dishes, and later discovered that they could cook hot dogs in front of them.)

Few people lose sleep worrying about their brain boiling. For most, the real fear is cancer. Its precursors, atomic transformations and genetic mutations, require the ionizing power of short waves, and this, unfortunately, is a UV specialty. (An ion is an atom that has lost one or more of its electrons, thereby acquiring an electrical charge.) About 1 percent of the light hitting a sunbather is UV, which translates into a million trillion photons per second. All are capable of altering DNA. Epidemiologists have found that every 1-percent increase in lifetime UV exposure produces a 1-percent hike in skin cancer incidence. Obviously the bottom line is—avoid it.

Easier said than done. Hiding under a beach umbrella isn't good enough: One-third of UV scatters in the atmosphere and comes at you sideways. This is because light-scattering increases exponentially as wavelength shortens. Distant mountains appear blue because blue light is bouncing among the air molecules between you and the mountain; imagine far more UV doing the same thing at the same time. (Conversely, red light scatters very little, which is why photographs taken through a red filter yield exceptionally sharp images of distant scenes; none of the hazy bouncing blue that occupies the intervening air gets through the filter to make the landscape indistinct. It's a trick used by photographers for more than a century; and the ultra-sharp landscape won't even surrender its red-filter secret if the image is shot in black-and-white. It also explains why amber sunglasses, which also block blue light, can make the world seem clearer.) Due to this strong atmospheric scattering, half the UV reaching you comes not straight from the Sun but from the sky itself, from its invisible UV brightness. That's why sitting in the shade is inadequate protection against sunburn if you're exposed to a big swath of sky.

Or snow, or other reflectors. Water generally absorbs UV when it's calm but reflects it when it's rough or ripply; thus a lakeside picnic is far safer on a calm day than it is on a breezy one. Sand reflects just 12 percent of UV when dry and an insignificant 5 percent when wet. Likewise, vegetation absorbs nearly all UV; no green plant reflects more than 10 percent. So your surroundings strongly influence your likelihood of burning.

A hazy day screens away half the incoming UV, which is why desert dwellers face a far higher risk than those in muggy tropical locales. When the air feels very humid, the time needed to tan or burn is doubled.

A cotton undershirt blocks 90 percent of the harmful rays no matter how ugly it is. Window glass (single-pane) does the same, transmitting just 12 percent. Cut that even more (in half) for standard double-pane. If you tan in one hour outdoors, it will take fifteen hours behind a typical window, a futile activity. To stay safe, do your sunbathing behind glass.

Why is glass opaque to UV? Fascinating stuff. Ultraviolet waves exactly match the electron resonance period of glass, creating high-amplitude vibrations. These vibrations jostle adjacent atoms so much that they can't settle down quickly enough to create new photons to continue through the pane. So UV striking glass produces heat rather than light.

The time of day and the season of the year determine how much ultraviolet arrives to do mischief on Earth's surface. As the Sun loses elevation, its light weakens plus has to pass through more blocking air. The UV environment becomes so low during the cold, low-Sun months that sitting behind glass in winter—say, in a greenhouse—you'd need at least 160 hours of steady noontime exposure to burn. To tan, you'd have to sunbathe nonstop for several days, an unlikely activity for the laziest of cats.

Even outdoors it's hard to tan or burn in winter, because of the Sun's low altitude. But since snow reflects between 75 and 90 percent of the ultraviolet striking it, you can negate the safety of November–January outdoor activity by going on a skiing vacation. Schussing later in the season is even worse; snow in March (when the Sun is much higher) produces a burn three times faster than in December.

From May through July the Sun is so high between 11:00 A.M. and 3:00 P.M. that anyone with normal skin will burn in less than ninety minutes. You do have ample safe summer hours before 9:00 AM and after 5:00 PM. During those seven early and late daylight hours, the UV-A intensity is more than halved and UV-B is chopped by at least 80 percent. The Sun may then look and feel strong, and skin will temporarily redden because infrared wavelengths reach you at nearly full strength. But you won't burn. So the safest outdoor summer activity—no hat or lotion required—would be a late afternoon picnic in a meadow: Grass reflects just 3 percent of UV.

Travel readily alters UV exposure, which is greatest in the tropics, of course, because of the Sun's high elevation. Greatest, too, in the Antarctic, because of the hole in that region's protective ozone layer. But simply going to places of higher elevation is nearly as dangerous. Every 1,000-foot climb raises UV intensity by 4 percent. That's why Denver's UV is 20 percent stronger than Philadelphia's, though both sit at the same latitude. An even bigger factor in UV exposure is the presence or absence of clouds—which is why Australians, with their sunny climate, have such a high rate of skin cancer compared with most Americans. Essentially, a heavily overcast day means no UV. ("Heavily" is the key word; thick dark clouds block everything, while a high thin cirrus layer lets almost all the bad photons through.)

Ozone, an odd molecule made of three oxygen atoms, is the frontline barrier for blocking dangerous UV-B. It's a skimpy, fragile shield: If all the atmospheric ozone settled at the surface, the layer would only be as thick as two pennies. Certain chemicals, like chlorofluorocarbons (CFCs) and halon (used in refrigeration and fire extinguishers, respectively, and now being phased out or banned outright), are such effective ozone destroyers that UV levels have grown markedly in the past twenty years. In temperate parts of the globe, the greatest increase (as much as 5 percent per decade) happened above latitude thirty-five—which unfortunately includes Europe, Canada, and the United States north of Texas.

It's worse at even higher latitudes—much worse. Because of its extremely cold temperatures and strong circumpolar winds that isolate the region, the ozone loss in the Antarctic, first noticed in the 1970s, has progressed to where ozone is now completely gone in winter between seven and twelve miles up, where ozone is normally strongest. The situation hasn't deteriorated since the hole reached its worst in 1993 only because ozone is currently zero in that region and can't get any lower. (Happily, the North Pole does not have a correspondingly severe problem, because its winds are different and its temperatures are much warmer.)

Although continued production of ozone-destroying chemicals has been virtually banned since the Montreal protocol of 1987, the long-lived molecules will wreak damage for years to come. Researchers expect UV levels to keep rising (albeit at a slower rate than before) before they finally start to fall, perhaps around the year 2015. Meanwhile NASA physicists will rely on monitoring the UV situation with satellites that include a series of Total Ozone Mapping Spectrometer spacecraft named TOMS. With these, experts can assess predictions that UV-blocking ozone will return to normal levels by 2050.

There's not much the rest of us can do except be ultracareful about ultraviolet. And, unlike mad dogs and Englishmen, stay out of the midday sun.

· Chapter 17 ·

Nature and Numbers, Rivers and Pi

Is the universe fundamentally complex beyond human imagining, like the tax code? Or does it operate by laws that are at heart wonderfully simple? Does all of nature point toward an underlying oneness, as theorists assume when seeking a Grand Unified Theory that combines all forms of matter and energy?

Consider two mathematical revelations, one easy, one complex, both of this past century. They nicely illustrate this "elegantly simple versus intricate" business.

First, that giddy academic headline in 1995: A previously obscure Princeton mathematician, Andrew Wiles, solved a three-century-old problem called Fermat's Last Theorem. The mathematical world cheered, and Wiles grew rich and famous overnight.

Pierre de Fermat (1601–1665), like many before him, had looked at the well-known Pythagorean theorem (the sum of the squares of two sides of a right triangle equals the hypotenuse squared) and wondered why the cubes of the sides never equal the cube of the hypotenuse. Fermat said he'd found a proof why no other power but the square of the sides can work. However, he didn't commit his proof to paper, and mathematicians ever since have tried to figure out what he had in mind.

Enter eight years of arduous labor by Wiles. His solution, hailed as the culmination of centuries of effort, takes skilled mathematicians days to confirm. It's 200 pages long, and (being far short of genius) I cannot follow even a typical line of it:

"... Thus if W (k) is the ring of Witt vectors of k, A is to be a complete Noetherian local W (k) algebra with residue field k and maximum ideal m, and a deformation (p) is just a strict equivalence class of homomorphisms ;:Gal (Qe/Q) - Gl2 (A) such that pmod m=Po, two such homomorphisms being called strictly equivalent if one can be brought to the other by conjunction of an element of ker: Gl2(A) - Gl2 (k) ...." And so on.

Imagine 200 pages of that! Obviously Fermat never did prove his theorem, at least not by the mind-paralyzing methods Wiles used. No wonder he procrastinated.

At any rate, this was one of math's holy grails for hundreds of years— something that most people would just shrug off as neither important nor particularly interesting. The point? There is nothing elegant about a proof so labored, forced, and contrary to the simplicity of its origin (the Pythagorean theorem, learned by everyone in junior high) that it can be grasped only by the smallest percentage of human brains.

Opposed to that, consider our second numerical notion, one much easier to understand. It's something Einstein liked to share with others, but he was not the first to note it: The average river's length, following all its curves, when compared with the straight-line distance from its source to its terminus at the ocean—yields pi! In other words, rivers tend to be 3.14 times longer than a straight line from their origin to their endpoint.

This is generally true because any bend in a river causes the water on the outside of the curve to flow faster, accelerating erosion and deepening the curve. But the river's bends do not exceed more than a semicircle, on average, because they would then bend back on themselves to close off and create an oxbow lake. In flatter terrain, rivers take paths curved differently from those of mountain rivers, but the average of all these balance out to pi.

Clear and understandable relationships like this application of pi to the natural world is what turns most people on. We all admire simple connections between numbers and nature, especially when we can understand them.

That Einstein should embrace this physical link with math is hardly surprising: He believed that the cosmos operates by laws and structures beautiful in and of themselves, and as logical as math. The great physicist always felt that a harmony filled the cosmos, an intelligent perfection he used as a test of his theories. His formulae not only worked, they *looked* elegant, and resonated with balance and even simplicity. (Well, not always. The field equations of general relativity, describing how objects distort and then move through curved spacetime, are so arduous that even NASA chooses not to use them when calculating spacecraft trajectories.) But his special relativity theory, and his famous $e=mc^2$, which demonstrates a stone-simple relationship between matter and energy, are wonderfully comprehensible.

Few would go so far as to insist that such elegance be a litmus test of an idea's validity. But the cosmos actually does appear to march in sync with mathematics. Which is why we sometimes speak of Newton as having discovered rather than invented calculus. And when Pythagoras uncovered his famous proof, he sacrificed a hundred oxen to express his gratitude to the gods. Such insight and inarguable certainty, he believed, must have divine rather than human origins.

It's still common to look for harmonies in nature, such as numbers that align with the physical world. The greatest and most obvious involves the only two disks in our sky: the Sun and the Moon. One is almost exactly 400 times larger than the other but is also 400 times farther away. Result: They each appear the same size. (More on the Moon in chapter 22).

A few centuries ago, when the search for other straightforward harmonies between math and the cosmos were all the rage, a real gem was uncovered by the German astronomer Daniel Titius in 1766 and popularized by his countryman Johann Bode a few years later. Eventually this layout became known as Bode's law, with its name justly revised in the past two decades as Bode-Titius. Nowadays we don't call it a law because it has no physical basis for existing at all. But back then, because the coincidence was so incredibly compelling, everyone assumed that a scientific justification for that arrangement would eventually be found. It never was. Here's what Bode-Titius is all about:

Start with the sequence 0, 3, 6, 12, 24 . . . in which each number (after zero) doubles the preceding one. Then add 4 to each, divide by 10, and you've got .4, .7, 1.0, 1.6, 2.8, 5.2, 10.0, and so on. What's so interesting about this? Just compare it with the distances from the Sun to each planet (expressed in standard Earth-Sun spans called astronomical units, or AUs):

PLANET	BODE-TITIUS	ACTUAL DISTANCES IN AUS
Mercury	.4	.4
Venus	.7	.7
Earth	1.0	1.0
Mars	1.6	1.52
Asteroids	2.8	2.8
Jupiter	5.2	5.2
Saturn	10	9.54
Uranus	19.6	19.18
Pluto	38.8	39.44

Yes, this omits Neptune from the picture, which at 30 AU doesn't fit into the scheme at all, but Neptune wasn't discovered until much later, in 1846. The discovery of Uranus in 1781, however, followed by that of the first asteroids twenty years later, fit the pattern perfectly—further "proof" that the heavens did indeed conform to the tidy numbering scheme. The arrangement is now regarded as a giant albeit remarkable coincidence. And an instructive one. For it teaches us to be cautious when linking events or phenomena with number schemes.

Those who enjoy finding solutions to the seemingly insoluble will appreciate the following, which is certainly one of the finest puzzles ever. It may seem unsolvable at first reading, but it yields a crisply gratifying, unambiguous answer. While it may take an intelligent reader an hour of thought, you'll regret giving up too quickly. Great pleasure derives from deciphering it. Solve this and you're ready to find math/nature links wherever you turn. (For those short of time or patience, the solution can be found in the appendix.)

Two women meet after many years. The first asks, "How old are your three daughters?" The second woman says, "The product of their ages is thirty-six."

FIRST WOMAN: But that's not enough information.

SECOND WOMAN: Well, the sum of their ages is the same number as the room you and I shared in college.

FIRST WOMAN: That still is not enough.

SECOND WOMAN: The oldest one has blue eyes.

FIRST WOMAN: Thank you, now I know the girls' ages.
What are the ages of the second woman's three daughters?

If you think that's tough, pity the theorists who try to make logical sense of conditions (for example, black holes) where all our math and physics simply dead-end! They may well dream of floating down a lazy river, sustained by nothing more than a piece of pi.

Part II

What's Going On Out There?

Chapter 18

Now You See Us,
Now You Don't

In most respects, our planet is an atypical region of orderly calm in the universe's turbulence; violence and calamity are par for the cosmic course. There are worlds like the Moon that have not experienced anything exciting for the past 3 billion years. There are also places like Jupiter, where unimaginable lightning and lethal radiation are routine. Luckily we lie somewhere in the middle. If our environment were inert enough to be perfectly safe, it would be too insipid, too lacking in complexity, for evolution to have taken place—DNA would never have woven its spirochete lanyards. Conversely, a bubbling porridge of unrelenting upheaval would quickly break apart complex molecules. Again, no life. So we *had* to live in a fairly stable planet that nonetheless springs nasty and inconsistent surprises upon us with diabolic irregularity. We can't really complain; it's the way things must be.

There's no limit to the close shaves and what-ifs that would have prevented us from ever existing in the first place. If any of the universe's four forces were to be tweaked in any way, you're not here to read this.

This very perfection influences modern thinking, since an explanation must be found for all the suspiciously fortuitous physical parameters. Is human life in some sense the *raison d'être* of the universe? How can we explain why we live in such a highly unlikely reality?

Not a problem, said Princeton physicist Robert Dicke in papers written in the sixties and elaborated upon by Brandon Carter in 1974. Their idea, which Carter dubbed the anthropic principle, stated that what we can expect to observe "must be restricted by the conditions necessary for our presence as observers." Put another way, if gravity were a hair stronger or the Big Bang a sliver weaker—and therefore the universe's lifespan significantly shorter—we couldn't be here to think about it. Since we're here, the universe had to be the way it is and therefore isn't unlikely at all. Case closed. By this reasoning, it's not at all remarkable that we live in a narrow comfort zone, on a planet a particular distance from a particular sort of star and bathed in a rich environment of elements produced by the death of earlier generations of suns. The seemingly fortuitous, suspiciously specific locale, circumstance, and time frame are just what's needed to produce life. If we're here, then this is what we must find around us.

Such reasoning is known as the weak version of the anthropic principle. The strong version, which is far more controversial because some think it strays from science into philosophy, says that the universe *must* have those properties that allow life to develop within it, because it was obviously "designed" with the goal of generating and sustaining observers. Going even further, the great Princeton theoretical physicist John Wheeler (who coined the term "black hole") introduced what some have called the participatory anthropic principle: Observers are necessary to bring the universe into existence. This idea arose to satisfy some of the strange observer-required phenomena of quantum theory, although Wheeler himself has recently backtracked a bit on the issue.

Many in the astronomical community guardedly embrace one version or another of the anthropic principle. Says astronomer Alex Filippenko of the University of California at Berkeley, "I like the weak anthropic principle; it has some predictive value. After all, small changes to seemingly boring properties of the universe could have easily produced a universe in which nobody would have been around to be bored."

Critics, however, wonder whether the principle is no more than circular reasoning, a facile way of squirming out of explaining the towering peculiarities of the physical universe. Philosopher John Leslie, in his 1996 book *Universes,* points out that a man in front of a firing squad of a hundred riflemen would be pretty surprised to find that every bullet missed him. He could certainly say to himself, "Of course they all missed; that makes perfect sense, otherwise I wouldn't be here to wonder why they all missed." But anyone in his or her right mind is going to want to know how such an unlikely event occurred.

One theory he might entertain is the "many worlds" explanation (see chapter 30) where numerous other simultaneous universes exist in which he did get shot; he, however, is aware only of the one in which he escapes.

Earth: Great While It Lasted

There is no sure safety in the heavens above. On some scale, we are continually witnessing worlds in collision. Every moment, a meteor punctures the sky over some spot on the planet. As we saw in chapter 1, a building in North America is struck every year and a half on average, and a big noisy intruder comes in once every 1,000 years or so. Even the word "disaster" comes from the old astrological belief in an extraterrestrial cause for catastrophes (dis: failing; aster: star).

Suppose we get lucky and manage to avoid the next five expected asteroids of Permian dimensions, each of which could leave the biosphere struggling to recover and blast away our earthly scene like dust in a car wash. Even if our planet defies all laws of chance and evades the next truly big ones, or we build the anti-incoming-asteroid defense system periodically discussed but not yet funded, our world will survive more or less intact for no more than another 1.1 billion years.

That's a non-negotiable limit because the Sun is steadily growing brighter. It's now calculated that it will become 10 percent hotter in just over a billion years. Ten percent hotter may sound like something we could get used to, sort of like moving from Boston to Atlanta. Unfortunately, that seemingly modest increment will be more than enough to evaporate the oceans and create a worldwide thick overcast that will trap heat—a runaway greenhouse effect. Equilibrium won't arrive until thermometers everywhere register a uniform 700°F. Sizzle city. Final curtain. Blackout.

We know that life began here some 3.9 billion years ago, and it would be nice if we were merely in middle age, with another 4 billion left. But no: We're 80 percent through our allotted custodianship of this planet. We—earthly life in all its varied forms—are like a sixty-year-old, a galactic senior citizen.

If humans survive to that final time when everything must perish, will we display the sanguine resignation typical of old folk who have seen it all and are ready to move on? Or will we find innovative ways to pack up our terrestrial belongings and relocate to the nearest hospitable planet, which is the now standard sci-fi vision? When you leave your childhood you're in exile forever, but when we finally relocate we may fully genetically adapt to the new world, and *it* may seem our true home, with Earth a vestigial imprint of our roots. But why speculate? Language itself will be unrecognizable in just a few hundred millennia; our very ways of thinking, our three-pound brains and the elaborate circuitry through which our neuro-electric currents connect, will be dynamically changed. No question asked today will generate an answer fathomable in a few million years—long before contract-canceling disasters are likely to happen. (Stop me if you've heard this one: A student nervously asks his astronomy professor, "When did you say life on Earth will be destroyed?" "A billion years from now." "Oh, thank goodness, I thought you said a *million!*")

The Universe: Don't Worry, Be Happy

And if uneasiness about the death of Earth a mere billion years from now seems unduly anxietal, what about the current fascination with the fate of *everything*? One of the hottest cosmological topics of the past twenty years has centered around the universe's expansion. How quickly is the cosmos blowing itself apart? The answer necessitates using a unit of volume such as a cubic light-year, a truly staggering measure. It would take a speeding bullet 380,000 years to traverse one of its edges, and the space within its boundaries is beyond imagining. But if you can grasp it, and also seize the enormity of a trillion, you can appreciate the universe's expansion with a single factoid: At the current rate, the cosmos continually grows 40 trillion cubic light-years larger every second.

Will it go on expanding forever?

Maybe it's a human eccentricity to spin our wheels by considering issues that cannot affect us or our descendants. Nevertheless, many resources are being devoted to this issue even if it will be at least 5 billion years (and more probably 10 or 15 billion) before it makes the slightest difference whether the universe stops expanding or keeps right on going.

If ever there was an epic drama that has no personal impact, that is conceptual rather than practical, this is it—a distant issue as irrelevant as a pod of dolphins debating whether the White House lawn should be mowed on Tuesdays or Thursdays.

Which do *you* root for, a closed universe where everything that exists crushes itself down to the size of a grape? Or its alternative—an inflated, ever-expanding cosmos of emptiness and absolute cold? Perhaps the seductive thrill that disasters evoke runs so deep in our psyche that we're keen on imagining future ones because we know we won't be there to revel in them. Coexisting with our very human need for security is a wild, mischievous, lunatic urge to knock down the sand castle, to scream into the galactic storm: "Go ahead! Hit me with your best shot!"

And given enough time, the universe will oblige.

Chapter 19

Oops

Success rarely has startling or enlightening consequences. It is the bizarre, unpredictable, or sadly tragic blunders that are often more illuminating—and certainly more interesting than when projects go as planned. Right this moment, somewhere on our spinning globe, people are experiencing food shortages, water shortages, and certainly money shortages, but there's never a shortage of goof-ups. Governmental, corporate, personal, minor to monumental, there may be only one way to do something right, but an almost infinite number of ways to mess things up.

Many spectacular examples of Murphy's law are so unlikely that no fiction writer could have imagined them. This "law" originated with U.S. Army Captain Edward Aloysius Murphy Jr., an engineer at Edwards Air Force Base in the California desert, who in 1949 noticed that the experimental rocket sled about to whiz off with Army Major John Stapp had had its wiring harness sensors installed backwards by a technician. Because the harness's design allowed the device to be inserted either correctly or 180° the other way around, Murphy concluded that "If there is any way to perform a task incorrectly, someone, someday, will perform it incorrectly in just that way." It is not known when (or by whom) Murphy's observation was shortened to its present form: *Anything that can go wrong will go wrong.*

And, of course, it does. For instance, many mistakes over the years have been due to a lack of punctuality. But punctuation? In 1962, the Venus-bound spacecraft *Mariner 1* had to be destroyed because it strayed off course during launch, a goof caused by the omission of a hyphen in the mission's computer programming. This single punctuation mark cost taxpayers $18 million. (Some have claimed that it was a missing comma, but the error in the FORTRAN DO-loop was a "bar," similar to a hyphen.)

Another spacecraft, the Russian *Phobos 1,* en route to Mars in 1988, was lost at the command of a programmer who sent a flawed computer instruction. The mishap didn't surprise many people: The Russians have a perfect failure record of attempts to fly by, orbit, or land on the Red Planet. Twenty tries, zero complete successes. *Phobos 2* managed to arrive in Martian orbit the following January, but then its first computer died, the second malfunctioned sporadically, and—too late—the Russians realized that the spacecraft wouldn't accept commands from the single remaining processor. A decade later, in 1999, U.S. programmers, as if in a perverse game of one-upsmanship, managed to outdo the Russians when the Mars Climate Orbiter plunged to its doom in the Martian atmosphere after being fed those conflicting units.

What's strange is that Mars, closer and easier to reach than most of the other planets, has hosted such an abysmal string of failed missions. By contrast, NASA has scored a perfect five-for-five with missions to Jupiter (two *Pioneers,* two *Voyagers,* and *Galileo*), five-for-five with Saturn (*Pioneers, Voyagers,* and, it appears, *Cassini*), and two-for-two with the *Voyager* visits to Uranus and Neptune.

Goof-ups in space science are usually costly, often spectacular, and always instructive. The archetypal U.S. foul-up was the Hubble Space Telescope fiasco in 1991. Its 94-inch mirror was the most precisely polished ever made—except that it was polished to the wrong shape.

Never before have so many involved in the space sciences laughed and cried at the same time. The blunder involved a billion-dollar snafu by the prestigious optical firm of Perkin-Elmer, in Danbury, Connecticut, which had won the contract to produce the world's finest mirror.

The idea behind Hubble was obvious. Unlike ground telescopes operating below miles of soupy, churning atmosphere, an instrument in the vacuum of space could finally capture the dreamlike beauty of nebulae. It would be like stumbling upon your first optometrist after an ageless myopia. Hubble would pinpoint quasars at the edge of the universe and observe planets with "being there" clarity. But gold-and-cobalt galaxies are more than merely pretty;

Hubble's instrumentation could resolve in them the cosmological issues that have vexed human brains ever since they first grew large enough to be tormented. It could supply some of the desperately needed hard data that would explode with a loud *whoosh* into the conceptual vacuum created by theoreticians. It would satisfy everybody.

But could we really pull it off? Could we defy our own bumbling, uneven history and have the raisins without the bran? We are, after all, the species that create magnificent Mayan pyramids and then shove virgins from their summits. The ones who carry life-saving penicillin into one jungle while slash-burning others to produce more hamburgers. (And those responsible for countless foul-ups in small towns everywhere, like the Woodstock, New York, plumber who connected a hot-water line to a toilet and didn't catch the mistake until its owner phoned to report the steamy, unsolicited elegance.)

Hardly a history suggesting that we could do anything right the first time—especially a project of unique complexity.

And we didn't. Hubble's misshapen primary mirror threw a cream pie into the face of NASA, which was blamed for improperly supervising the contractor. But NASA didn't deserve the entire blame. Turns out that when testing the optical surface, Perkin-Elmer used an expensive state-of-the-art device called a reflective null corrector, which told them that the mirror was perfect. But when they double-checked the shape with an older instrument called a refractive null corrector, that analysis revealed a serious deviation. (In this case, "serious" meant that the mirror's edges were off by one-fiftieth the thickness of a human hair.)

Now, what is the procedure when two tests give contradictory results? Check the alignments of both testers? Perform a third test? Here's what Perkin-Elmer did: nothing. They decided that the older instrument must be wrong. So a telescope was blasted into orbit with a defect so egregious that any amateur telescope maker would have caught it. It cost two years, an intricate repair mission, and over $100 million to correct Hubble's spherical aberration. And the whole mess could be traced to a single optical element in the fancy reflective null corrector, which itself had been misaligned by the thickness of a dime.

Hubble's now-corrected optics continue to amaze us all with otherworldly images of such beauty and utility that its initial woes now seem like the labor pains accompanying any birth. But pragmatists may wonder: If such an expensive, high-profile project can screw up, can't anything? Could we ever assure future colonists signing up to live on a moon of Jupiter that they won't freeze into ice sculptures because someone forgot a bolt or washer?

No one who heard them can forget the screams of the three *Apollo 1* astronauts when a simple spark, a minor "short," turned their pure-oxygen environment into a blinding inferno that melted the spacesuits off their bodies during a 1967 practice on the launchpad. Turns out, the craft's oxygen environment wouldn't have been quite so dangerous in space, where a fire's combustion products hover in place and tend to smother the flame. But in Earth's gravity, where heat and smoke rise and new air is pulled in to feed the fire, the launchpad practice made a pure-oxygen environment a terrible idea. Ah, hindsight!

To this day, people still speculate about the four-minute-long anguish of the 1986 *Challenger* astronauts, that handsome, winsome crew, as they fell almost nine miles, alive and fully conscious until their still-intact cabin smashed into the sea at 207 mph. "Space disaster" again made headlines in 2003, when the oldest shuttle, *Columbia,* broke apart at 207,000 feet during re-entry. The fiery fragments and the dark rain of thousands of jagged pieces of debris over five states made it history's most observed aeronautical accident. (Oddly, all seventeen American astronauts who suffered fatal accidents did so during a single black-lined week, between January 27 and February 1.)

The Russians have fared even worse. Cosmonaut Vladimir Komarov's *Soyuz 1* impacted the ground at 400 miles per hour after its parachutes failed to open properly. Reportedly Komarov wept while speaking to his wife from orbit, knowing how badly things were going and the probable outcome that lay minutes away.

Boris Volynov, the first Jewish cosmonaut, had nothing but trouble but managed to survive—barely. His 1969 *Soyuz 5* mission suffered multiple problems, and things turned critical during re-entry, when his equipment module failed to fully separate from the descent module. Blocking the heat shield, it made the craft unstable, and *Soyuz 5* entered Earth's atmosphere tumbling out of control. When it finally stabilized, the wrong section (the thin-skinned nose) was facing forward and the hapless craft started to burn up. The situation had reached the irreparable point, with the craft beginning to break apart, when the equipment module ripped off and the descent module flipped over, heat shield now properly forward. But his troubles weren't over: When the parachutes emerged, they only partially deployed. Volynov could do nothing but wait to die in the high-speed landing. But when the dust cleared from the brutal impact he was still alive, though most of his front teeth had been knocked out. Meanwhile the Russians had lost track of his position and had no idea where he'd come down. When rescuers finally arrived, Volynov had already walked to the nearest village, a toothless intruder attempting to convince the peasants that he'd just come from space.

The all-time rocketry botch job, the event most suggesting brain impairment, was committed by Marshall Mitrofan Nedelin, commander of Russia's strategic missile forces. On October 24, 1960, at the famous Baikonur launch site (named for a distant town to confuse Western intelligence), Nedelin gave orders seated on a folding chair a few yards from the giant rocket about to be test fired. Over workers' objections, he insisted on saving time by having a faulty electrical circuit fixed on the spot without draining the fuel tanks that had just been filled. An upper stage ignited with a roar while still attached to the enormous fuel-laden main stage. The resulting fireball killed ninety-two people and injured another thirty, more than the casualties of all other rocket and spacecraft mishaps combined. Exactly three years later an oxygen leak caused another explosion that killed eight people at the same site. From that day forward, engineers and workers have refused to work at Baikonur on October 24.

For some astronauts, bad mishaps could have been even worse. One Mercury astronaut (Gus Grissom, who later died in the *Apollo 1* accident) very nearly drowned after exiting his just-landed capsule, which flooded and sank because he'd prematurely blown the hatch. And Neil Armstrong lived to go to the Moon only because he scarcely survived a practice mishap a few years earlier. The lander device malfunctioned a hundred feet above the ground; Armstrong ejected and was catapulted just barely high enough for his chute to open in time. The margin between life and death had been less than two seconds.

These and a myriad of inevitable future foul-ups must be factored into the equation when we contemplate human space travel or any novel enterprise in a hostile environment. Most of us live in the comfortable lap of western civilization. Television, supermarkets, soccer games, school plays, mortgage payments. Neither agony nor ecstasy. Our errors in judgment at work or home rarely cost us more than time, money, or reputation; our appliances and vehicles are specifically designed so that a single error is not usually fatal. (Engineers know it's hard to make anything *totally* idiot-proof because idiots are both stubborn and ingenious.) Few everyday objects (with the exception of motorcycles and some power tools) are fundamentally unforgiving. But in the milieu of test pilots and astronauts, a solitary error is enough to do you in.

Even before the February 2003 *Columbia* tragedy, NASA insiders had pegged the odds of a catastrophic space shuttle accident at somewhere between 1 in 50 to 1 in 200—per flight. Ride the shuttle three times, as some have, and you've faced as much as a 1-in-17 chance of dying a spectacular death. Sadly, those odds will not likely improve with the modifications engineers make in the wake of each disaster. The shuttle's complexity, speed, and operating environment offer too many ways in which it can fail. Future space vehicles are not likely to be much safer, either.

Forever imperfect, we cannot expect to leave behind this very human trait to travel up a trouble-free boulevard to the stars. Those brave enough to volunteer will encounter far more than their share of blood-curdling blunders and narrow escapes, as they take us on our odysseys to other worlds.

Chapter 20

The Discovery That
Shook the World

Nothing—black holes and supernovae included—fuels excitement like the discovery of a new planet. Maybe because we live on one, the announcement hits close to home. But perhaps our ongoing interest in things planetary owes its origin to a collective residual memory of the most astonishing scientific announcement *of all time*—the finding of Uranus on March 13, 1781.

Today that giant aquamarine dollop of hydrogen inspires as much awe as a cheeseburger. On our lists of the most exciting places to explore, Uranus fits somewhere near Dubuque. But imagine yourself a learned, urban resident of the late eighteenth century. The age of science and reason is well advanced. New lands have been explored and colonized. The Industrial Revolution has begun to transform long-established modes of life. In an era of discoveries, what could arouse the jaded sophisticate?

Here's what: The realization that some fundamental aspect of reality had suddenly changed forever.

The existence of five and only five extraterrestrial planets had been taken for granted since the dawn of sky watching by ancestors dressed in furry underwear. It was a certainty. Search for others? No more reason than for scientists today to seek evidence that the Sun is hollow and filled with steaming broccoli.

Some things just seem totally safe from revision. Would we expect modern science to alter its opinion about the basic function of the human heart, or announce that dogs are in continual contact with aliens from Alpha

Centauri? Citizens of the mid-1700s were equally confident that our solar system contained one Earth, one Sun, and five planets. Period. Seven was the "number of perfection," and there were seven days of the week to honor the sky's celestial bodies. The utterly unexpected discovery of a giant new planet was more than a revelation: It was a bombshell that jolted everyone's security. If there were more planets, then everything became questionable. It was a humbling and frightening experience.

After the initial astonishment, self-flagellation began. Why hadn't anyone seen this before? Uranus is dimly but clearly visible to the naked eye, and changes position among the background stars. How had it escaped the notice of all the supposedly keen-eyed Arab desert dwellers who had named the stars in the first place? Where were the Chinese observers who had chronicled the supernovae when the Europeans were sleeping, during the Dark Ages? And what about the first 180 years of telescopic observation: Uranus is positively brilliant through *any* telescope. It had indeed been previously spotted and catalogued; the problem was that nobody had noticed that the little "star" waltzed across the sky. Why had it taken so long for someone to notice its motion and its green unstarlike disk? How could this colossal oversight have happened?

Although the discoverer, German-born British astronomer William Herschel, suggested calling the new world Georgium Sidus after his royal patron, the public's first impulse was to name it after its finder, a fitting reward that nonetheless aroused a storm of protest. Not because "Herschel" lacks resonance. If it sounds ridiculous to us now, it's no worse than Uranus, the most mispronounced celestial name after Betelgeuse (few correctly say YUR-an-us or BET'l-juice). Herschel and "George" both lost out because the tradition of naming planets for Roman gods was too established to be ignored. So Uranus eventually prevailed and everyone got used to it, more or less. And Herschel did receive his due: The symbol for Uranus is a stylized lower case "h."

The discovery prompted a near-frantic search for yet more planets, a celestial gold rush that didn't pan out for a couple of generations, when the arduous hunt for various theorized Planet X's yielded Neptune, found in 1846. Galileo's drawings show that he charted Neptune more than 200 years before its official discovery. Oddly, the great observer failed to recognize the slowly moving dot for what it was.

Pluto's discovery in 1930 made headlines, but the world had been primed for the finding by numerous reports about its suspected existence. (The first two letters of its name honor the man who most fervently believed in and searched for it and at whose observatory it was found—Percival Lowell.) Most astronomers now agree that the ninth "planet" is probably the largest example of a new class of tiny, cometlike ice chunks that inhabit the outer solar system. The sighting in 2002 of another Plutolike object a billion miles farther out (tentatively given the bet-you-can't-pronounce-this name of Quauar—that's KWA-war) has further diminished Pluto's planet status. From Pluto, incidentally, the Sun is not just a bright star but a point of light 300 times brighter than the full moon: too dazzling and dangerous to look at! If Pluto had an earthlike atmosphere, the Sun's brilliance would even give it a bright blue sky.

There just aren't many other examples of jaw-dropping discoveries as unsettling to entire populations as Uranus was. New planets beyond our solar system, detected in the last decade, have so far not been imaged, only inferred from their gravitational influence on their parent stars (and in one case by a slight change in the parent star's brightness). Their existence had long been expected by astronomers, so they were far from surprising.

Quasars were puzzling, violent, distant objects of unknown characteristics when first observed in 1966 but proved to be merely the explosive cores of ordinary young galaxies. They made a few headlines but never became topic A at the office cooler.

Penicillin was an unexpected discovery, but bread molds had been used in folk medicine in various countries for centuries. New elements (helium wasn't found until the late nineteenth century) were attention grabbers but were mostly inert substances that affected neither the lives nor the opinions of the average citizen. Even the airplane, exciting to watch, was an evolutionary rather than revolutionary leap. And while the past century has brought enormous technological advances, few have made the world gasp. The laser? It's everywhere now, reading supermarket bar codes, delivering music and movies on CDs and DVDs, and enabling eyesight improvements, but its introduction didn't stun society or instantly change the average person's mind or life. There's no denying that the first automobile ushered in great lifestyle alterations, but no sudden shake-ups there, either. (Cars were initially regarded as toys, then hailed as saviors of the environment! The stench, cobblestone din, and density of buzzing flies accompanying the crowded population of horses and carts in nineteenth-century cities made the automobile initially seem like an ecological redeemer.)

Yes, the first earth satellite, launched by the Soviet Union on October 4, 1957, commanded universal attention, but the race to put a satellite in orbit had been going on for some time, and the only real surprise was that another country beat us to it.

Despite (or maybe because of) this whirlwind technological era, it's hard to recall a scientific event or discovery that has shaken us to our underwear in a single dramatic moment—not the computer, not the advent of television, not even sending people into space. With the possible exception of the A-bomb, what most reshaped our daily lives snuck up on us; nothing leapt out and yelled *Boo!* Scientifically savvy laypeople may indeed have been astonished when an entirely new state of matter—the Bose-Einstein condensate—was announced in 1995, but since it can exist only at temperatures within a fraction of a degree of absolute zero it will not likely yield practical applications for a long time to come. Cold fusion commanded headlines but proved to be a bust. In the 1980s the cosmology crowd was dazzled by the news that the universe is structured like a sponge, with vast empty spaces surrounded by curving sheets of galaxy clusters. And stunned again by the discovery in 2000 that the expansion of the universe is accelerating instead of slowing down. But unfortunately such revelations are apprehended by only a small fraction of the public.

Even genuinely unexpected and off-the-wall scientific discoveries hardly make the ten o'clock news. No, to duplicate the shock of Uranus's discovery we'd have to stumble upon something that would shatter cherished, long-held beliefs, instantly revealing that we had been dead wrong about some universally accepted tenet of reality. But as science and technology advance with exponential rapidity, our capacity for astonishment shrinks to near zero. And Uranus, whose sudden appearance in our collective consciousness was so traumatic, has become, in our time, a comic's deliberately mispronounced joke.

Ah, but to experience total intellectual humility, just once!

To have been alive in 1781!

Chapter 21

The Forbidden Light

One of the strangest items in the cosmic inventory is also among the most overlooked: the curious Crab Nebula and the astonishing neutron star that hovers at its heart.

Neutron stars don't get much mention these days, despite being fairly new to us—having been discovered only in the sixties. To the public, they don't seem as morbidly fascinating as black holes or as capable of fully warping spacetime. The media shy away from runner-ups and wannabes. A pity. It's like ignoring a lizard-man sitting next to you at the movies just because you saw a pterodactyl in the parking lot.

This story really started before dawn on July 4, 1054, when a dazzling star abruptly appeared among the winter constellations, not far from Orion. As bright as Venus, it blazed for a year and was duly noted by observers in China and duly ignored in the West where alterations in the heavens were incompatible with the prevailing theology. Now, nearly a millennium later, every telescope user can see the result of that destroyed sun: the Crab Nebula.

Talk about weirdness! Consider the odd glow visible even through binoculars, next to the leftmost horn tip of Taurus the Bull. This radiance has a palpably alien quality, and for good reason. It's not starlight. Not sunlight. Not reflected light. Its origin is neither heat nor nuclear energy. The glow of the Crab Nebula comes from an exotic phenomenon bearing an alluring name: forbidden radiation.

Forbidden radiation looks unearthly to us because it's so unfamiliar. Even our most high-tech labs can't produce it—hence the name. It appears when isolated atoms have been excited by synchrotron radiation, which is generated when frenzied electrons are whipped to near light speed by unimaginably powerful magnetic fields. If you're picturing the kind of magnetism that makes compass needles wobble, think again. The field near the Crab is 900 billion times stronger than Earth's. It blasts electrons into violent spiraling geysers that defile any atoms in their way. It would rip that compass right out of your hand.

Dwelling in a harder vacuum than we know how to create, the region's oxygen atoms are ionized into a unique metastable state where they squirm uncomfortably for hours. Finally their extra energy escapes, in maneuvers so improbable they never occur in our own cosmic neighborhood. *Zap!* Floods of green forbidden light stream off into the universe—a strange energy whose turquoise hue is glimpsed through large amateur telescopes of sixteen-inch aperture or more. But this is just the cloak surrounding the inner sanctum of singular strangeness. Now we get to the heart of the nebula, the phenomenal object creating that magnetic field and forbidden radiation in the first place. The Crab pulsar.

At first, "pulsar" seems simple enough to define. It's a tiny, hard neutron star whose magnetic poles sweep past our line of sight with each rotation. Neutron stars, in turn, are massive suns with the weight of two to four of our own, or about a million planet Earths. The heavy star's immense gravity overwhelms the quantum pressure that normally keeps subatomic particles apart, so its entire million-mile-wide body collapses into a sphere so tiny it could fit on a small coastal island. Picture a star on Martha's Vineyard.

Like a lighthouse, neutron stars can deliver quick bursts of energy with every turn, and because they spin very quickly—up to nearly 1,000 times a second—there's a steady series of ultrafast flashes in a wide spectrum of wavelengths. In the case of the Crab, the spin rate is thirty-three times a second. If Earth happens to lie in a direction where it can receive these pulses, the neutron star is called a pulsar. Their light looks steady because nobody can differentiate so many flashes per second. (The human "flicker fusion threshold" is twenty flashes per second. Fewer than that—as in old 16mm movies—and we see flickers. More than that, as in today's movies flashing at twenty-four frames per second, and the image looks steady.)

The collapsed star's intense magnetic field acts as a brake and is slowing the wildly spinning Crab pulsar while-U-wait. By the year 4000, it will whirl just seventeen times a second—a flickering finally noticeable through backyard telescopes.

This insanely spinning object, this spherical remnant of the unfortunate sun that blew itself to Kingdom Come on the Fourth of July 1,000 years ago, actually exploded a millennium before the first Pyramids. Some 5,000 light-years away, the subatomic wreckage traveled for fifty centuries before slamming into us.

It would have been amazing to witness the shattered star sucked inward by its own gravity. What could be more dramatic than watching a star abruptly deflate like a punctured balloon? The material of a nearly a million planet Earths suddenly packed itself into a globe smaller than New York City. Laws of physics that normally keep objects apart but fail in places like supernovae and subway cars allowed this sun, bigger than our own, to collapse down to a radius of only six miles.

Its atomic contents were fused into a compressed swarm of neutrons, with some packed leftover protons and electrons elbowing in vain for a little breathing room. "Dense" or "hard" are pathetic understatements. The average density of Crab material is the same as a sugar cube containing every American automobile. Imagine a one-centimeter cube with the weight of 100 million cars. Or, how's this: The star's present density would be equaled by an aircraft carrier crushed down to the size of the ball in a ball-point pen. Except here there isn't merely a tiny speck of the stuff but a sphere twelve miles across containing nothing but.

Amazingly, this is the same density found in every atomic nucleus in your body—and in everything else. So in a way the Crab is one giant neutron, an enormous atom fragment one two-hundredth the width of the Moon.

We usually think of star surfaces as gassy and unsupportive. (The Sun's vaporous visible photosphere is much less dense than water.) But the Crab's surface is something else. Unimaginable gravity has forced its broken atom fragments into a kind of lattice, its exterior skin a spherical glasslike structure 100,000 trillion times harder than steel. It doesn't need to be insured. You can't damage it.

A half mile beneath its impenetrable crust the star turns to fluid yet becomes even more dense. This layer boasts a curious super-slipperiness that lets the interior effortlessly spin at a different rate than the star's surface. A few miles down at the center lies a mystery: Our present science is limited to guesses; nobody knows what's there.

The only visible object that could in theory surpass this famous crushed superball would be a quark star. While quarks exist in protons and neutrons only as groups of three, they might be able to pack themselves singly to produce a Guinness-World-Record density. In 2002, researchers announced finding exactly such an object, which made the front page of the *New York Times,* though the claim was soon rejected by almost the entire astrophysical community, which cited better explanations for the data.

So the Crab retains its status as the densest object viewable through backyard telescopes, and remains on the top-ten list of the strangest things one can ever see. Even its color is an illusion: In addition to the otherworldly green glow of the nebula, the light of the pulsar itself turns ruddier as it speeds across the universe, redshifted simply from the effort of fighting its way out of the sticky gravitational epoxy and onward toward our eyes and brains.

Worthy of more than the media's fifteen-minute allotment of fame.

Chapter 22

Our Nearest Neighbor

Deeper even than the passion of poets, the vows of sweethearts and the demented howling of wolves is our enchantment with the Moon. It permeates our legend and lore, influences weather and transportation, and maybe even our emotions. Not to mention billions of tax dollars spent sending people and robot spacecraft to pay house calls.

All because the newly formed Earth was clobbered by a stray Mars-sized planet about 4 billion years ago, according to the most plausible theory. Myriad molten fragments of Earth blew off into space; some rained back down, some escaped our gravitational purview forever, and others coalesced into a sphere, like falling mist merging into droplets. This newborn moon was very close to Earth at first, probably only a third of its current distance, and occupied a large and dramatic portion of the sky. But it never achieved the critical seven-mile-per-second speed needed for a complete escape. If it had, we would have joined Mercury and Venus as lonely moonless worlds. Instead, the Moon settled into an elliptical orbit, where it's been spiraling away ever since at the breakneck speed of 1.3 inches a year. (To be accurate, the surfaces of the Moon and Earth only grow an inch farther apart annually. That's because the highest point on our planet, the 29,035-foot summit of Mount Everest, itself rises moonward by a quarter-inch a year, shoved upward by the northerly drift of the Indo-Australian plate on which it rides.)

Each remnant of the collision, Earth and Moon, suffers permanent withdrawal symptoms of that Siamese-twin separation. Both are forever bound by immense tidal tugs from the other. Earth experiences these as a daily surface deformation of nearly a foot: The very ground rises toward and falls away from the Moon as we rotate. More familiarly, our oceans rise and fall three feet worldwide, producing the average five-foot coastal tides. A newly discovered atmospheric tide is yet another Earth-Moon link, and may be the reason that the weather is slightly rainier and cloudier during the full moon than at other phases. The effects go on and on, and include oddities noticed at least since the time of the ancient Greeks. For example, the full moon is neither duller at the edges nor slightly brighter at its center, both of which would convey the impression of dimensionality; instead it has the look of a flat disk, a set designer's cutout prop. No matter how often we see it, it still seems odd, as if the full moon were painted onto the sky.

Strangely, too—and unlike every other major moon in the solar system—the Moon does not circle the equator. Its path ignores Earth's tilt and instead moves in the same plane as that of the planets orbiting the Sun.

And if the Moon influences Earth, the reverse is eighty-one times greater: Earth affects the Moon enormously. That little piece of escaped terrestrial real estate—eighty-one times less massive than Earth—suffers a tug from our world that makes terrestrial tides seem piddling: The Moon undergoes tidal stresses some twenty times greater than ours.

The contest is all the more imbalanced since the Moon's density is only half that of Earth's due to its sandy, silicate composition, carved from Earth's mantle rather than our metal-rich core. (Imagine a tug of war between Arnold Schwarzenegger and Woody Allen.) The side facing Earth suffers the greatest tug, continually braked by Earth's tidal pull. This is what caused its rotation to freeze like hardening cement so that the same face of the Moon always appears before us like a frame from a jammed projector. Some people think it's an amazing coincidence that the Moon's rotation and revolution are the same. It's no coincidence at all. Like a choreographed chorus line, every satellite of every planet shows the same synchronized spin—an invariable effect produced wherever a heavy body's tidal effect influences a lighter nearby neighbor. Creatures on any planet would, like us, see one face alone of their moon or moons, with the other hemisphere forever hidden. As logical as the explanation may be, the result remains fascinatingly odd, as if we always saw the same profile of a horse and rider regardless of whether they were on the near or far side of the race track. The tedium is barely relieved when an effect called libration allows us to see tiny slivers around the edges of the Moon's opposite hemisphere. There simply is no other body in the universe that shows us just one of its sides. Mercury comes closest, by exposing the same hemisphere to us on every alternate revolution around the Sun.

The lunar hemisphere forever unseen from Earth was a mystery until October 4, 1959, when a Soviet spacecraft flew there and transmitted its images. Everyone had assumed that the far hemisphere would look much like the one we see. Wrong. Again. The unfamiliar hemisphere upset all predictions. It is entirely different. On the near side, numerous cold, frozen, ancient flows of lava drown a myriad of even older craters. The flows created maria (seas)—enormous dark zones, the most conspicuous lunar features seen by the unaided eye. The opposite side has almost no maria and instead displays countless craters made during the first 100 million years of lunar life. (That's because the relatively smaller tidal force on that hemisphere resulted in far fewer volcanoes and crater-covering lava flows.)

All of the Apollo missions landed exclusively on the near side because radio communication with Earth is blocked by the Moon itself, not because the far side is "dark." Despite the unfortunate phrase "dark side of the Moon" in poor science fiction and even poorer science writing, every lunar neighborhood experiences both day and night at half-month intervals. Yet the notion of a world with one side forever in shadow is so seductive that a megahit album like Pink Floyd's *Dark Side of the Moon* will surely not be the last we'll hear of that mythical nonexistent realm.

Seeing the Moon

Strangely, the Moon is one of the few perfectly round objects viewable in the natural world. In a diameter that is 2,160 miles across, its equatorial bulge of just four miles is imperceptible. Having seen that sphere countless times throughout our lives, you'd think that the familiar "Man in the Moon" pattern of moon blotches would be imprinted in our minds. Not so. We easily recognize the features of hundreds of movie stars but not the face of our lifelong nocturnal companion. That's why filmmakers can palm off faked, reversed, or upside down Moons. Nobody notices. (We're routinely oblivious of everyday things; for instance, can you recall which way you turn a doorknob—clockwise or its opposite—in order to enter a room?)

Our nearest neighbor lacks color in both character and fact. Geologically dead, it is one of the few totally gray objects in the known universe; the visiting astronauts wasted color film on its uniformly monochrome surface. Also, like the ragged hermit who turns out to be a millionaire, the real Moon differs markedly from the harsh landscape of jagged mountains that appear through binoculars or telescopes. In reality it is a land of gently rolling hills covered with deep and incredibly fine powder. The austere lunar terrain is blanketed with silky baker's-flour softness.

For just one night out of thirty, our satellite becomes beloved by vampires and despised by astronomers—the full phase. During that time, the critical shadowing that defines craters and mountains in sharp relief disappears and the rest of the sky is overwhelmed by moonlight. There would seem no good reason to be attracted to such a featureless disk, far less interesting than either the crescent or the football-shaped gibbous phases. Yet this is the phase most favored by artists and lovers, the one that garners the lion's share of lore and myth, the one credited with supposed powers and abilities never attributed to any other. What hip lunatic or werewolf would emerge with the first quarter? As the most popular of lunar guises, the full moon demands exploration.

Its basic properties are simple and should be known by everyone (but aren't): The full moon always rises as the Sun sets. Sets as the Sun rises. Is highest when the Sun is lowest (at midnight, or 1:00 A.M. during Daylight Saving Time). Is the only phase not seen in daytime. Is the only phase that's out all night long. And it's the only possible phase that can be eclipsed.

When full, the Moon becomes 30,000 times brighter than Sirius, the night's most luminous star. Still, the Moon is a poor light reflector: Its terrain reflects only a paltry 8 to 13 percent of incoming sunlight, making it as dark as asphalt. By comparison, Earth's albedo (the percentage of light we reflect) is about 35, while Venus bounces off a whopping 76. If a Moon-sized Venus lived as close to us, we'd be bathed in light seven times brighter than the light of the full moon—enough to give us a blue sky whenever this miniature Venus was full.

The lunar drabness itself is a distinction. In all the observable universe, only testimonial speeches are duller. The only reason the full moon looks bright is its contrast with the night sky. Even if some ambitious real estate developer were to pave its entire surface with asphalt and turn it into an enormous parking lot, it wouldn't get any darker. If it were seen against a sunlit bedsheet instead of a dark background, it would appear nearly black! This can't be stressed enough: The Moon seems white only because it's seen against a darker background, and our retinas, like a camera's exposure meter, automatically compensate. (Skeptical? Tape a small dark-gray paper disk to a sheet of white paper, then place an identical disk on a black background. The disk's brightness will magically change, depending on the color behind it.)

The Moon's true luminosity is surprisingly difficult to pin down subjectively. When asked to compare moonlight with sunlight, few make a reasonable appraisal. One problem is that we never see Sun and full Moon side by side. The Sun sets; there is a gradual darkening of the sky while our eyes slowly shift into their night (scotopic) vision mode. Pupils dilate and photochemical alterations change the ASA film speed of our retinas. Since no direct comparisons are possible, it's not surprising that the 120 respondents to an informal survey conducted by the author were grossly off the mark. They guessed sunlight to be 50 to 200 times brighter than full moonlight. The correct answer: The Sun is 450,000 times more luminous!

There's mystery to the moon's brightness. One might reasonably assume that a full moon is twice as bright as a half-moon. It's actually eleven times brighter. Half its maximum luminosity is reached just 2.4 days before full phase and then it brightens almost explosively. The explanation for this rapid increase in brilliance is twofold. First, innumerable little shadows cast by lunar pebbles and sand grains when the Sun is low in the lunar sky all disappear when the Sun is overhead, as it is at the full phase. The second is simply the bouncing of sunlight straight back at the source, like a movie screen. We see this light only at full phase, when the Sun is directly behind us. This latter effect also explains the "flat moon" phenomenon.

Another surprise is how wrong most people are when estimating the full moon's apparent size. Asked to put a value on its dimensions, most guess that a line of thirty to fifty full moons would stretch from the horizon to the zenith. The true figure? It would take 180 full moons to fill that span. And if 180 Moons can span just 90°, the Moon must appear a minuscule 0.5° across. To completely pack the sky, 105,050 Moons would be needed. Its disk is far smaller than commonly imagined.

Time and Tide

At full and again at new moon, unusually strong spring tides nourish the popular lunar association with magical powers. Of course, lunar roundness or brightness has nothing to do with it; rather, the tides merely respond to the Sun and Moon positioned in a straight line and acting in concert. That's in contrast to first and last quarter phases, when solar and lunar tidal pulls partially cancel each other out because Sun and Moon are at right angles to us. Moreover, the Moon does not pull on water or the oceans any more than it pulls on your socks. It is only Earth's large diameter that does the trick: The lunar tug on our closer side is different from the tug experienced by the other, 8,000 miles farther away. This difference is not what *produces* the tidal effect, it *is* the tidal effect, and it acts on Earth as a whole, not on any of its constituent parts, like you or me.

As for lunar phases, greater powers do not accrue to a fully lit Moon any more than a friend standing fully bathed by the Sun has any more "power" than when he or she is half in shade. Still, in many dumb-and-dumber ways, the full moon is credited with influencing earthly affairs.

A connection between the full moon and Earth's weather was not pinned down until two investigations in the 1960s graphed a number of meteorological parameters over several years, revealing a slight 29.5-day periodicity that matches the Moon's synodic period of phases. Later, in 1990, the full moon was linked to a reliable periodic temperature increase in the lowest four miles of the troposphere (the air nearest the ground), which experiences a global monthly warming of 0.02°C (or about four-hundredths of a degree Fahrenheit) during the period from five to eight days after the full phase. This tiny monthly temperature rise is still not understood but may possibly be caused by the Moon's hot daytime surface throwing infrared heat our way like a radiator, or by some atmospheric interaction with oceanic spring tides, or perhaps even by lunar-induced atmospheric tides. The rainfall/cloudiness link is so weak that it shows up only statistically; nonetheless, there is a slightly greater chance of the full moon being hidden by clouds than any other phase.

Yon Inconstant Moon

Another rug pull-out in this slippery world: You can't count on a full moon every 29.5 days. That oft-published interval varies greatly. Recalling that the full phase is defined as when the Moon is most nearly opposite the Sun, we'd expect that if Earth's orbit were circular and its orbital speed therefore uniform, the full moon would reliably repeat in the same time interval. But it doesn't play that way. Earth's orbit is an ellipse that brings us as near as 91.5 million miles from the Sun and as far as 94 million. By Kepler's second law of motion, Earth moves faster when nearest the Sun, and the direction toward the Sun then changes most rapidly. Therefore, so does the direction opposite the Sun, where the full moon is located. These factors change the interval from one full moon to the next, a variance from 28.5 days to nearly 31.

Here's Looking at You

Nearly forty years after the first Moon landing, earthbound observers still yearn to relate to adventures in the sky. Observatory visitors often ask hopefully, "Could today's telescopes have seen the astronauts on the Moon?" It's fun to fathom our optical limits, and it's easy to grasp just what we can and cannot perceive from Earth. If we do someday establish a Moon colony, could we see it through Hubble, Keck, or Palomar, or perhaps even with amateur backyard instruments?

The answer: A good hobbyist telescope boasts about one-second-of-arc resolution, meaning it could detect a golf ball six miles away. Impressive—and Hubble, up in space, performs twenty times better. The Moon appears some 2,000 arcseconds wide and by happy coincidence is 2,160 miles in diameter, so a mile on the Moon appears about one arcsecond in size. Lunar features smaller than a mile are therefore close to hopeless through earthly telescopes, while Hubble's 0.05 arcsecond resolution could just make out one or two city blocks on the Moon. Obviously, nothing we brought there was that big. Even the lunar landing module would have been much too small for Hubble to see, had the space telescope existed then. But Hubble might see giant factory-sized structures, if Lunar City ever gets a building permit.

Moon Gold

Data collected from all six of the successful *Apollo* landings on the Moon between 1969 and 1972 confirmed the assessment of the earlier, unmanned lunar probes: It's geologically dead. Its 500°F temperature range from day to night, its utter lack of air and water, and the paucity of valuable minerals or resources made it thoroughly unappealing as any kind of semipermanent home. Had its rocks not been found to be uniformly anhydrous (meaning you can't even squeeze water out of them chemically) and had humans not needed ample (and at eight pounds per gallon, weighty) amounts of it for survival, those pioneering voyages might indeed have been "one small step" toward permanent settlement. But nobody seemed anxious to replay the adventure, and more ambitious launches motivated by dreams of future colonies were ruled out as exercises in fantastically expensive water hauling.

That is, until twenty years later, when the United States Army's *Clementine* space probe (strange in itself, since the army was never in the Moon-exploring business) found the signature of water ice in permanently shadowed craters at the lunar poles. That momentous and utterly unexpected 1996 discovery, confirmed in 1998, changed everything. In a large crater on the Moon's south pole, forever shaded from the Sun like some vampire lair, lies a lake of ice. This single frozen oasis makes all the difference. Instead of needing to haul every drop of water at hellacious cost, we can now count on a supply awaiting us at the lunar south pole, a fact that alters forever the course of human ambitions and possibilities. Once considered a worthless, uninteresting, dry wasteland, the Moon is now a worthless, uninteresting wasteland with a single huge freezerful of ice. If we ever want to make it a waystop, a jumping-off point to more distant places, it has just gone from unsuitable to possible. Like the old wooden railroad water towers that serviced steam locomotives, the Moon may function as a water station for interplanetary shuttles.

Water—the true lunar gold—is needed not just for drinking but for disassociating into its component hydrogen and oxygen, which make an excellent rocket fuel that already propels the main engine of the space shuttle. Yet that's apparently not quite good enough; NASA has no plans to return. The next human steps on the moon will probably be those of Chinese, on their way to achieving their stated intention of reaching Mars—yet another market for made-in-China exports.

Chapter 23

Voodoo Moon

The Moon does influence us. But its powers scarcely resemble the strange lunar abilities fostered by tradition, legend, and popular belief. Obviously, the Moon's in-your-face presence as the night's dominant companion has resulted in a vast store of folklore. But can it really drive you mad? Influence plant growth? Affect moods and menstrual cycles? Cause accidents? Stimulate births?

Almost everyone who works in the maternity wing of a hospital subscribes to this belief. Echoing her colleagues, one veteran nurse in an upstate New York hospital told me, "I've been here twenty years, and I don't just think so, I *know* so: We certainly do see more births here around the time of full moon. Ask anybody."

In one way, she's correct. Medical professionals do commonly perceive more births at the full moon. And this is curious and worthy of study—because it's wrong. An exhaustive review of fifty years of twenty-one Moon/birth studies, published in the journal *Psychological Reports* (Vol. 65, 923-4) concluded that there is no connection between human birth and lunar phase. More recently, astronomer Daniel Caton of Appalachian State University analyzed 70 million birth records from the U.S. National Center for Health Statistics. Result: No connection exists.

The strong *perception* of a relationship between births and the Moon is a separate phenomenon, one that explains much about lunar folklore. Whenever a busy pace of births coincides with a full moon, several of the hospital staff will typically say, "The nursery's jammed and more coming in. It figures, there's a full moon out there!" But when the full moon lands on a slow night, nobody comments, since no unusual activity triggers discussion. Over time, the more-births-at-the-full-moon idea is reinforced, but it is never weakened. Indeed, many will still call the Moon full when it is out-of-round. And on hectic cloudy nights, when the Moon is neither full nor visible, some will still exclaim, "It must be a full moon!" The reinforcement is a one-way street, and no one involved is aware of the process.

With regard to other alleged lunar connections—to hyperactivity in children, automobile accidents, moodiness—it is wise to remember that people who look for associations usually find them. The only way to know whether a perceived relationship is real or not is through careful and objective analysis of reliable data—not always easy to do, because popular books and magazine articles are often biased or riddled with bogus statistics. A typical failing is to cite a study indicating that crime is linked with the Moon without bothering to mention the twenty-five studies that found no connection at all. Some make statements about lunar gravity "pulling on body fluids," thereby revealing total ignorance about how the Moon's tidal effect works. Or they attribute behavior to the Moon's "magnetism," when in fact the Moon has no magnetic field. The topic of lunar influence floats in a sea of misinformation. In reality, statistical studies have shattered the idea of lunar linkage to homicides, alcoholism, fire alarms, domestic violence, suicide, major disasters, prison violence, emergency room admissions, assaults, traffic accidents, crisis calls to police stations, epilepsy, and sleepwalking—among other civil and personal disturbances.

But just as the Moon's "not guilty" verdict seems assured, new evidence is submitted to the jury. Several other patterns are indeed tied with the Moon. The following is a brief summary of the best evidence for the Moon's actual influences.

Animal Behavior

Animals active during daylight hours generally come into heat in seasonal rather than lunar patterns, but a number of nocturnal animals do indeed have lunar-linked biological rhythms. Some are related not to tides or moon phase but simply to when the Moon is visible. For example, a study found that rats are least active when the Moon is highest in the sky and most energetic when the Moon is below the horizon. Natural selection would favor rats that remained hidden during bright nights, when predators such as owls could more readily find them.

Interestingly, the metabolic rhythms in some animals match neither the twenty-four-hour day nor the twenty-five-hour lunar cycle. Birds kept from experiencing daylight, for example, exhibit a strange twenty-three-hour biological clock that follows neither Sun nor Moon.

Marine Life

A great number of ocean organisms have tide-related life cycles, which of course are synchronized with lunar rhythms. The greatest influence of the Moon is probably in the intertidal zones, the muddy marshes and endless beaches whose environment so radically changes with the twice-daily ebb and flow of the sea. Here life depends on burrowing into the wet sand at clockwork intervals or finding other ways to survive periods of exposure. Algae, oysters, barnacles, snails, worms, mollusks, crabs, and many others, and the life upon which they feed—and their own predators, such as gulls— probably represent the biota most attuned to the position and phases of our nearest neighbor in space. Even when fiddler crabs are removed from an ocean environment, their periods of peak activity will continue to advance by fifty minutes a day, matching the lunar cycle of tides.

Human Biology

The most likely link between us and the Moon involves our circadian rhythms, or our biological clock—which is preset to regulate numerous cycles, from sleep to appetite. Like those of most other animals, our clock follows a simple daily twenty-four-hour activity pattern. Yet when people are placed in a lightproof room and live without outside references to the day/night cycle, their bodies undergo a curious change: Sleep and metabolic patterns advance daily by very nearly the average fifty-minute time that the Moon rises later each day. Such a 24.9-hour pattern has also been reported in a blind person living in normal society.

Given the importance of psychogenic factors in the production of both physical and mental illness, one would assume that a delicately balanced person aware of the full moon's classical association with "lunacy" might exhibit increased aberrant behavior. However, while a few studies have suggested a lunar tie to some aspects of behavior, a recent review of a hundred such studies by educational psychologist Ivan W. Kelly and two colleagues at the University of Saskatchewan found no statistically significant link between the Moon and mental illness.

Numerous other studies, including one that covered 4,190 suicides in California's Sacramento County over a fifty-eight-year period, found no connection between suicides and lunar phase. A survey of 32,000 calls to a crisis center also showed no link with the Moon, full or otherwise. Of the eight published studies on the association between lunar phase and mental-hospital admission, only three found such a connection—but with contradictory lunar phases! And even these refuted the classic association with the full moon: The two largest (25,000 psychiatric-emergency-room visits and 18,500 psychiatric-hospital admissions) showed the *lowest* rate during full moon.

Crime

An analysis of fourteen years of homicides in Dade County, Florida, by Arnold Leiber, a psychologist at the University of Miami, indicated a link between murder and lunar periodicity. But when the study was later subjected to standard statistical review by University of California astrophysicist George Abell, the link was found to be spurious. An examination of homicides in New York City shows the same pattern of irrelevance to lunar phase. And though one survey of 34,000 violent crimes did show an increase around full moon, a much larger study involving 311,000 calls for police assistance found no correlation with lunar phase.

Fertility and Menstrual cycles

In the United States and several other countries, more births happen in the month of September than at other times of the year. Either humans experience a slight seasonal reproductive cycle, or else (more likely) this blip is related to the preceding long, dark, cold December or the increased leisure time around the holidays. Beyond this seasonal connection, many studies have looked for a link between our reproductive cycle and the Moon. Any such association, even a subtle one, would be doubly interesting, because the average length of pregnancy, 266 days, happens to match nine synodic months of 265.8 days. Thus a time-of-birth link would simultaneously suggest a time-of-fertility connection with the Moon's phases. (On average, the Moon's phase at birth is the same as it was at conception: If you were born under a full moon, chances are you were conceived during a full moon, too.)

This ancient and popularly accepted link remains controversial. Over 4,000 years ago, the Babylonians believed women to be fertile according to the Moon, though their lack of accurate timekeeping must have made the Moon's near-correspondence with the menstrual cycle seem like a perfect match. Despite such a long tradition, no more than 22 percent of all women have menstrual cycles that match either the Moon's 27.3-day sidereal period (time of revolution) or its 29.5-day cycle of phases. If a woman does have her period on the day of full moon (or any other phase) it is rare for that to habitually and reliably repeat.

Even if the Moon does not causally match the female reproductive cycle at the present time, it may still have relevance to it. Many millennia ago, humans were probably more apt to venture out on bright than on dark nights. Perhaps our reproductive rhythms grew attuned to the recurrent times of bright nights, when it would make more biological sense to be fertile. Dates on dark moonless nights were more likely to end early when you or your boyfriend would be some creature's entrée. If so, the close match between the average menstrual cycle and the Moon's period of phases is no coincidence but rather a legacy of the many millennia when moonlight had real relevance to everyday life.

Arguing the other way, only the opossum has a similar menstrual cycle to humans. We'd then have to assume that only humans and opossums were chosen by nature to be linked to the Moon in this manner, with the rest of the animal kingdom exempt. (For example, the estrus cycle is twenty-one days for cows, twenty-five days for macaques, and thirty-seven days for chimpanzees.) A further powerful argument against the connection is that if menstruation is linked with the Moon then so, too, would time of fertility and—since human gestation averages exactly nine synodic months—the birthrate would also show a marked resonance with lunar phase, which it does not.

Taken together, the current data neither decisively proves nor disproves that the close matchup between synodic and menstrual cycles is coincidence rather than correspondence.

Planting by the Moon

In April 1794, the *Old Farmer's Almanac* counseled that "Wheat sown at this quarter of the moon is not subject to smutting." In the centuries since, the popularity of planting by the Moon has scarcely diminished. The Anthroposophist Rudolf Steiner advocated it in 1924. While some almanacs simply offer astrological or other irrational planting advice (a kind of gardening horoscope that has no scientific or factual basis), the subject of lunar links with plant growth has inspired only a small number of serious studies, with mixed results.

Most agronomists dismiss lunar links as folklore, pointing out that planting by the Moon began because the lunar calendar was the best way to know when the right season had arrived for various crops. The issue is confounded by "biodynamics," a nonscientific system that links plants, various Moon rhythms, and signs of the zodiac. This scheme was later disproved (as if it needed disproving) by studies conducted near Frankfurt by the biologist Hartmut Spiess. As to straightforward phase links with plant growth and yield, Spiess's controlled studies with winter rye planted at daily intervals found a slight lunar link with the early stages of growth—a link that vanished as the plants grew and ended up with "little effect on yield." One U.S. study found a tiny (1 percent) Moon-linked change in oxygen respiration in potatoes, but only in the fall. Other studies showed no effects at all. Any lunar connection with nonmarine plant growth would appear to be small at best.

Very little legitimate peer-reviewed research on Moon planting has been performed, to say nothing of duplicated. Cornell agronomist Robert Beyfuss found that folklore parading as fact pervades the gardening-advice literature. He points to the widely accepted practice of "companion planting" carrots and tomatoes side by side, which purportedly helps both to grow better. When he laboriously tracked all the many books and articles that repeated this "fact," he discovered that each had picked up the idea from earlier sources rather than from actual research papers. And it all traced back to a single German, Ehrenfried Pfieffer, who in the 1930s wrote that when he crushed tomato and carrot plants the extracts yielded colors that were "compatibly clear and bright."

Emotions

Why do some people feel romantic when the Moon is full? As we've seen, a bright Moon may have been the only nocturnal condition in which our distant ancestors felt safe enough to pay cave calls on the opposite sex. As far as other emotions are concerned, we should remember that the Moon's gravitational pull exerts no more force on a person than does a brick wall a few feet away. Lunar gravity cannot physically influence people's brains or glands, or sway a single molecule of liquid. But what about the Moon's *light*?

Beyond the fact that even full moonlight is a half-million times less bright than sunlight, we live in an era when artificial light makes moonlight superfluous; sadly, other sources of illumination routinely overshadow the glow from our nearest neighbor. Unlike sunlight—and its wintertime absence, which is known to profoundly affect at least 15 percent of the United States population with a depression-producing syndrome appropriately termed SAD (Seasonal Affective Disorder)—there is no known physical mechanism by which a human body would either be helped or harmed by the feeble light of the Moon. However, since the full moon has a slight correlation with increased rainfall and cloudy weather, and cloudiness is indeed linked with a change in mood, here at last is an actual (if extremely subtle) mechanism by which the Moon *can* influence emotion, at least statistically.

But Moon emotions may go much deeper than this. Many report that moonlight evokes an ineffable "feeling." Though such a description is not quantifiable, people inarguably have collective memories, instinctive familiarities that have been summoned for countless centuries. Even without invoking ancestral imprints, most people have conscious or subconscious childhood recollections of moonlit nights and carry such sensations through life. Like love and other imponderables, this unnamed impression is undeniable even if it cannot be scientifically demonstrated.

Finally, the Moon's mythical link with romance and courtship so pervades poetry and literature that most societies and individuals have a strong collective mindset in this direction. The Moon may well have an actual effect on mood and emotion stemming from sheer expectation and belief. In this sense, some of the Moon's more curious alleged "powers" may be valid—without even needing to be true.

Chapter 24

Space Frolics

The sixties and seventies were a wonderfully Moony time. The Marcels' "Blue Moon" was a big hit. The college fad was full-monty mooning. The Rev. Moon gave us the Moonies. Quarterback Warren Moon was the rising star. We *had* to go to the Moon.

And we did, in six landings that sent a dozen people hopping along its hot, fine, powdery surface during the Moon's early morning hours, chosen so that the astronauts would enjoy optimum lighting before the ground heated up excessively. So from Earth the Moon was never either new or full during the missions. There were other restrictions: They all landed on the hemisphere that eternally faces Earth, of course, on sites closer to the middle than to the lunar limb. All of those areas are simultaneously visible for one week each month—within four days of the full moon. The naked eye has no trouble finding the now famous Sea of Tranquillity, a dark blotch to the lower right of the center, where *Apollo 11* landed and celestial soil was first disturbed by human feet.

After three manned missions had circled the Moon, the first that landed was rife with glitches, starting with Neil Armstrong muffing his famous speech by leaving out one of its dozen words. Omitting the "a" in "That's one small step for [a] man" fudged the meaning, even if nobody seemed to care. (Despite the lack of any other truncated exchanges with Houston, Armstrong later insisted that he had indeed included the "a" but that it had been lost in transmission.) For the record, the actual first word from the Moon was the all-American "OK"—spoken not by Armstrong but by Buzz Aldrin, part of the first complete lunar utterance, "OK, engine stop." Only then did Armstrong say, "Houston, the *Eagle* has landed."

Aldrin, the second man to step down, felt the urine bag inside his boot break the moment he made contact with the lunar surface. *His* walk could have been characterized as "one small squish for man." Urine bags were among the items left behind on the lunar surface by all six missions. Just before liftoff, each crew would open the hatch one last time to toss them out, along with other paraphernalia not needed for the long flight home. The impact of these urine bags on the surface registered on the nearby seismometers planted by the crew, sending ground-vibration signals back to Earth. (In the category of trivia, let it be noted that these were the only times that pee has been officially recorded on earthquake monitoring equipment.) Another generally unpublished minor *Apollo 11* mishap involved the American flag, which had been planted on a mast with much difficulty and fanfare: it fell over, as if on cue, at lunar blastoff.

Such little-known facts aside, so much has been written about that initial landing that it's only fair to focus on the less celebrated missions, numbers 12 through 17. Today, few remember those wildly successful odysseys that spanned some of the most eventful years in history—and the people who flew them. Time for a brief airing of some curious incidents.

Apollo 12, with Alan Bean, Richard Gordon, and Pete Conrad, launched four months after that first Armstrong/Aldrin adventure. It got off to a frightening start. Thirty seconds after blastoff, the mighty Saturn 5 rocket was struck by lightning. All the breakers tripped, and the astronauts' hearts pounded as controllers watched their frantic EKGs on screens far below. Luckily, power was restored within seconds and the lander touched down four days later, just 200 yards from an old robot *Surveyor* spacecraft, as planned. On one of their walks, Bean and Conrad removed chunks of it to learn how thirty months on the Moon had affected its materials. When the pieces were analyzed later back on Earth—a shock: a living bacterium was found! A germ (alpha-hemolytic *Streptococcus mitis*) had not only apparently contaminated a piece of the spacecraft but had survived for two and a half years in the Moon's baking 250°F heat, -255°F cold, total vacuum, and utter dryness. No wonder it's so hard to shake a strep throat!

Apollo 12's series of unpredictable events continued even through its final moments. Upon splashdown, a heavy camera supposedly attached securely to a wall shook loose and smashed into Bean's head. (Yes, Bean had been beaned.) Blood gushed and the victim lay unmoving, wide-open eyes staring blindly. Said fellow astronaut Pete Conrad, "It knocked him cold. I thought he was dead." Because the crew knew they were going to be on global TV in just a matter of minutes, a giant Band-Aid was used to conceal the injury—which remained the worst suffered by an astronaut on any of the *Apollo* flights.

Four months later the next trip to the Moon eventually earned its own book (*Lost Moon,* by Jim Lovell and Jeffrey Kluger) and hit movie. Looking out their window, trapped helplessly between the Scylla of space and the Charybdis of their dwindling oxygen, James Lovell, Jack Swigert, and Fred Haise barely survived. Four days of shivering in dark, damp, 38°F cold in their T-shirts was compounded (for Haise) by raging fevers and a kidney infection; he lost fourteen pounds on the aborted mission. Venting waste overboard was impossible and it had to be stored in various imaginative places. Yet the greatest oddity of that craft's harrowing and ingenious fight to survive involved the numeral itself. By insisting on proceeding with a mission labeled "13" to demonstrate its disdain for superstition, NASA ended up adding considerably to the number of people suffering from triskaidekaphobia. *Apollo 13* was launched on April 11, and the tremendous oxygen tank explosion that blew out its service module's entire side occurred on the 13th, as if to taunt NASA even further.

Apollo 14 saw Edgar Mitchell (adopting the role of self-proclaimed psychic Edgar Cayce) carry out a mental telepathy experiment with four people back on Earth, using cards with special symbols. Result: Only one person performed slightly better than chance, scoring 51 correct answers out of 200. (Hardly impressive, since 40 hits would have been expected from mere guessing.) The three other subjects failed to attain even random results. Lesson learned: Don't rely on ESP communication when traveling between celestial bodies.

That same offbeat mission featured Alan Shepard pulling a couple of concealed golf balls from his pocket and hitting them with a makeshift club fashioned from the handle of a geology instrument. While he first claimed that the balls went "miles and miles," a later, more sober analysis downgraded the distances to 200-400 yards. Still respectable.

Apollo 15, 16, and *17* each had astronauts spending three days on the surface, tripling the first landing's duration. All used the lunar Rover (cruising speed: 7 mph) to drive dozens of miles—thus confirming the world's suspicions that Americans really cannot go anywhere without a car. The rover was equipped with an automated navigation system designed so that the vehicle could find its own way back to the spacecraft. While one might assume that the astronauts could simply retrace their forever-imprinted tracks in the thick dust, a complicated circuitous series of ruts would have made the task bewildering. Moreover, their waiting spacecraft quickly vanished from view over the lunar horizon, which was a mere mile away, and getting lost on the Moon would have been a serious error.

Even in the realm of the weird, who would guess that one of the *Apollo 15* astronauts would deliberately leave behind a human replica on the Moon's surface, like a voodoo doll? David Scott placed the sculpture in a small depression on the surface, as a salute to the fourteen American and Soviet astronauts who had lost their lives in their countries' space programs. Any aliens arriving on the Moon a billion years hence—when humans will certainly be long gone or unrecognizably evolved—may find that aluminum likeness and wonder: What is this thing?

They'll also discover the numerous officially sanctioned plaques bearing then-president Richard Nixon's name and signature (at the insistence of the White House). Perhaps they'll link the two. In any case, they might well puzzle over what incredible accomplishments this long-ago individual must have achieved in order to be so memorialized. What did this "Richard Nixon" do to have his name alone placed over and over on the erosion-free surface of the Moon, to survive for eons of time? They'd have no clue, because none of the late president's famous quotes (such as "I am not a crook") were included.

In 1972, the *Apollo 16* crew drove seventy stories up a mountainside for the view of their lives. Roaming about, they also discovered a strong local spot of magnetism on the otherwise magnetically lifeless lunar terrain, an occurrence reminiscent of the discovery of the mysterious slab in *2001: A Space Odyssey*. The magnetic spot is now believed to be the remains of an ancient impacted meteor. *Apollo 16* established the lunar speed record when the pair opened 'er up back down the mountainside at 11 mph.

In the early eighties, fifteen years after the first landing, the *Apollo 11* crew held a rare public reunion on NBC-TV's *Today Show*. Armstrong, a dedicated introvert who refuses to make public appearances despite many lucrative inducements, was asked if his first steps on the Moon—the event that immortalized him—represented the apex of his illustrious career. "No," he said. The actual peak had been "the point after touchdown, when Buzz and I shook hands without saying a word." On that eternally silent world, the taciturn, history-making astronaut most savored the moments of stillness after the engine had been silenced.

The Steamy Side of Space

The record for detecting signs of humans in space occurred during the *Apollo 13* near-disastrous mission, when a Canadian telescope saw the expanding shell of gas from its exploded oxygen tank 200,000 miles from Earth. It's not good news for observers trying to follow the action. When astronauts explored the Moon, they were never out of mind. But they were definitely, as they said in the sixties, out of sight.

This fact of being hidden from Earth's prying eyes has inspired much speculation about secretive activities beyond Earth's atmosphere. To date, many women have accompanied men in space, mostly on the shuttle. Well, did any couple turn up the heat in zero gravity? Insiders at NASA say, "Definitely." Although NASA officially denies that there has ever been the slightest hanky-panky, it's an open secret that so far about a dozen weightless male and female astronauts joined what NASA insiders call the hundred-mile club. The very first? This is a particularly touchy topic. However, a crew member of a mid-1980s shuttle flight had made it clear to a number of people that she intended to gain that historic distinction. One of the most obvious liaisons, these insiders say, occurred during the 1992 shuttle mission STS-47. An openly steamy affair had been going on between two of the scheduled astronauts aboard that space-lab mission, Jan Davis and Mark Lee, and they were married shortly before blastoff. NASA maintenance personnel later reported finding "clear evidence" of this only-too-human diversion (as they had in cleaning up the shuttle after several earlier flights).

Travels with Ivan

In 1998, NASA astronaut Andy Thomas spent four and a half months aboard the Russian *Mir* space station. While he recalls loving the Russians and particularly their distinctive space food, he found their cavalier attitude about safety distressing. A malfunction started a fire, which they were fortunately able to contain, and the acrid smoke lingered in the station for days, inhaled nonstop by the crew. When Thomas expressed concern about "whatever it was we were breathing," the Russian commander just laughed it off.

I asked Thomas, who later served aboard the International Space Station, what the most difficult or unusual aspect of life in space was. He said it was orientation. On the ISS, different modules are arrayed in varying directions; when you go through a doorway from one to the other, the direction toward the ceiling on one becomes the direction toward a wall on the other. He says that whenever he arrived at a doorway he had to pause for a few seconds to decide which way to turn his floating body "in order to make the floor the floor." With no gravitational up or down in space, everyday life would have been bewildering if the modules had not been deliberately arranged so that the astronauts could distinguish between floors, ceilings, and walls and consciously align themselves floorward throughout the day to perpetuate this psychologically necessary sensation. Thomas said that a common prank, played particularly on newly arrived astronauts, was for the others all to turn themselves upside down when the novice was momentarily out of the room, so that he or she would become disoriented upon arriving back at the doorway. Potato chips and M&Ms posed another problem; snack fragments commonly floated off in all directions, some hopelessly irretrievable.

NASA chief engineer Rob Landis reports that "mice and rats scurried everywhere" during his 1999 stint at the Korolev Control Center near Moscow, where U.S. astronauts and NASA personnel co-commanded the ISS with the Russians. "You didn't want to put your lunch or your candy bar down for a minute," he says. When NASA staff complained, the Russians in charge roared, "No problem! We'll poison them!" and soon the stench of dead animals rose from the floorboards. A year later, when Landis next returned, the scene was different but just as peculiar. Now cats were omnipresent, prowling, jumping up on consoles, running through the halls, lounging on nearly every unoccupied chair. The Russians had solved the rodent problem, but now the aroma of cat droppings filled the high-tech space center. Janitors perfunctorily cleaned things up by smearing the poop with their brooms, and there were stained floors throughout the building. Outside the center, another animal menace waited: packs of wild dogs. Russia has no humane society to deal with the numerous canines that have been abandoned in the region. "You didn't dare wait by yourself for the train back to Moscow," said Landis. Added another NASA engineer, "You'd never see this in Houston."

In fact, you'd never see a lot of Russian stuff in or above the United States. "There was far more alcohol aboard *Mir* and *Salyut* than anyone imagined," reported a NASA official in a position to know. One Russian space module was dedicated to smoking. Up in space they offered what no American airliner any longer has: a smoking section!

Future space vehicles, equipped and programmed to explore unimaginable distances, may have no need of astronauts. But it will always be good to remember, as in *Camelot,* the "one bright and shining moment" when a few men and women endowed with enormous courage and exceptional intelligence escaped from the constrictions of our planet. Those first tentative steps into space may have had their odd moments—but that has only made them more human.

Chapter 25

The Big Stink

Ah, sipping tea amid the garden scent of flowers while hearing dawn's symphony of songbirds! Life's full enjoyment enlists all the senses.

But not for astronomers. Their field is one of the few in which knowledge arrives via the single sense of sight. We cannot smell, hear, taste, or (except for meteorites and Moon rocks) touch the celestial objects of our fascination.

Birdwatchers listen as much as look. Gourmets taste and sniff. Astronomers simply stare—though many people imagine that they also *hear* the cosmos via radio telescopes. Movies perpetuate this myth by portraying radio astronomers wearing headsets, as if downloading the latest Kazaa music from Sagittarius. And if not headsets, then with ears cocked like beagles toward wall-mounted speakers, whose hisses and crackles suddenly snap everyone to attention by changing to an ominous drone. In actuality, radio waves are just another form of light (electromagnetic radiation), and radio telescopes cannot detect sounds of any kind.

Astronomy is a medium of light alone. And even here we're trapped within provincial boundaries. Our eyes are blind to all wavelengths except the narrow visible part of the spectrum, between 4,000 and 7,000 angstroms, where the Sun most strongly emits its energy. In a way, we view the universe only with the Sun's sanction: We're blind to gamma-rays, X rays, microwaves—everything that the Sun either barely emits or which is blocked by our atmosphere. Since our eyes evolved to see by daylight, we're all hardwired to scan the universe with a Sun bias. We're far more capable of perceiving sunlike objects (or whatever basks in sunlight, like planets) than anything else.

How might the universe be apprehended if we could employ all our other senses? Well, taste is probably ruled out: Who'd want to lick Pluto? Probably not a single researcher with access to any of the 842 pounds of lunar soil and rock ever thought, "What the heck, let's see how the Moon tastes." But wouldn't you pop a crunchy bit of lunar dust into your mouth just to know that you were the only person on Earth who had eaten Moon? (Having acquired a wonderful two-pound iron meteorite, an asteroid fragment that landed in Africa almost two centuries ago, I confess I once gave it a quick lick. No worse than cold pizza.)

What would really be fun is to play with some lunar regolith, the baby-powder-fine soil that completely covers the Moon's surface. The astronauts couldn't do that, of course, because they had to wear space gloves and keep the stuff under quarantine. Unfortunately, not enough has been assembled to fill a sandbox. Future lunar colonists would have to put up a protective bubble over a flat sealed piece of ground and pump it full of air, and only then could they romp around barefoot and let the kids roll in it.

Not much else in space cries out to be fondled. The solar system abounds with ice and cold rock, neither of which offers a novel tactile experience. Still, if you want to stroke a smooth piece of actual Moon rock, you can do it in Washington's Smithsonian Air and Space Museum.

More intriguing would be to *hear* the sounds of the solar system. The Sun is like a giant subwoofer, with enormous high-amplitude pulsations of various cycles. Some of those noises and beats could be heard all the way to Earth if there were an interplanetary medium to carry the sound waves. In fact, "heard" is an understatement; the Sun would be a deafening boombox, blaring combinations of regular deep tones punctuated by Latin-style flourishes. And the melody would change during eclipses, or at least alter in volume.

The vacuum of space insulates our ears from such potential uproar: A firecracker would be perfectly silent if it went off a few inches from an astronaut's head. That explains why the huge explosion aboard *Apollo 13* seemed so muffled; the percussion was largely directed outward into space. If there had been air around the ship, the astronauts might not have been so calm when they said, "Houston, we've had a problem." (Yes, that's the actual quote; they phrased it in the past tense—too optimistically, it turned out.)

Planets like Saturn which have constant fierce winds blowing over 500 miles per hour must experience a continuous terrifying roar. Jupiter's ultra-powerful lightning storms and Europa's ice sheets grinding into each other would also resound with extremely loud and complex noises. A musically creative astronomer might synthesize such effects as background accompaniment when visitors peer through a telescope.

For an even more enveloping experience, one could add the smells of the Jupiter system. Don't expect roses. This is a region even roaches would avoid, arguably the foulest-smelling place in the known universe outside of certain locker rooms. A health inspector would happily approve the sanitary, totally sterile conditions of Mercury and might even tolerate the sulfuric-acid smell of the Venusian clouds. But Jupiter? *Peee—uw!*

First you've got Io's volcanoes continually spewing sulfur fumes into space, bathing the entire region in the aroma of stink bombs. Then Jupiter itself has an atmosphere rife with methane (swamp gas) and ammonia. The place resembles a crowded dormitory where the dinner menu featured chili. No wonder the most likely place for life in that sector would be under the ice covering the oceans of Europa, where sensible sea-dwelling creatures would avoid all unpleasantness. (Actually, methane is odorless. It's the trace stuff, the hydrogen sulfide, that would get you.)

In most cases, unique local conditions would produce perceptual surprises. On Earth, 100-mph winds deliver sensory extremes in the form of earsplitting howls and brute tactile forces that can knock a person to the ground. Not so on the Red Planet. There, the very low air pressure makes winds of the same high velocity as soft and quiet as an earthly ten-mile-per-hour breeze. Moreover, sound barely travels on Mars, except through the ground itself. In the brief seconds before asphyxiating, a helmetless visitor would feel, smell, and hear very little.

Beyond the solar system, only the sense of sight would be generally useful. The universe is a place of isolated pockets of material in a vast matrix of emptiness; few if any sounds, smells, tastes, or tactile impressions are available. An exuberant hippie astronaut, attempting to experience the true nature of the universe by removing his spacesuit to feel it like it is, would be painfully disappointed. His body would swell grotesquely from lack of air pressure, while quickly freezing solid, with liquids in the eyes and mouth bubbling, boiling, and hardening at the same time. Even this swift biological violence would unfold in a theater of eerie absolute silence.

Our senses evolved specifically to access terrestrial conditions. That's why going anywhere else would punish us with extremes of either unpleasant overload or deprivation. So while it would be astronomically intriguing to expand our receptors beyond sight alone, it wouldn't be much fun.

All of which carries us back to the timeless reminder that there's no place like home. We may someday become interplanetary travelers, but only Earth can bring us to our senses.

Chapter 26

Constellation Consternation

Y
ou love the night and can recognize a few constellations. You'd like to learn more, but the task seems daunting: There are eighty-eight of them, the same number as piano keys, and the patterns are weird. Yet you can master the stars quick-and-easy and still appreciate their insanity.

First, a vital tip. You need to acknowledge that the constellations are a mess; there's no other way to put it. That this system of connect-the-dots is universally accepted is itself a puzzle. The handful of sensible patterns like Orion and Scorpius are only teasers, fostering the illusion that one can actually see mythological figures up there. In reality, the celestial landscape is crowded with faint Rorschach patterns that offer few associative possibilities. It's a Mad Hatter's night sky.

No sober person could construct a "Fox" out of the dim suns of the constellation Vulpecula. And Octans the Octant, where the south polar point sits, is so faint that it's not really there at all. For no good reason Aquarius is labeled "the Water Bearer" when it should more aptly be known as "Medley of Random Dots." Capricornus the Sea Goat (whatever *that* is) boasts the outline of New York State. So let's simplify this expedition by focusing on those rare islands of lucidity in the sea of chaos.

Once you've learned to recognize a particular constellation, you have to know your mythology in case your companion asks, "Who is that guy? What's a Centaur?" For example, summer and autumn evenings reveal the faint stars of Ophiuchus, The Serpent Bearer. Was there ever really such a profession? Well, maybe 2,500 years ago, but no employment agency has found work for one lately. Reptile portage has gone the way of the type-writer repair shop.

Modern occupations would not be much better. The chiropractor? The bus driver? No real improvement on what's already there. So the established order will endure, and the sky's archaic occupations, added to the ancients' penchant for immortalizing things like flies and snakes, guarantees that the firmament will remain forever bizarre.

Those of us living in the Northern Hemisphere at least see constellations named chiefly in classical times. Although most have alien names and con-fusing configurations, they've been screen-tested by a hundred human generations. When you travel near the equator or into the Southern Hemi-sphere, however, you're in never-never land. There the stars were named much more recently, by sixteenth- and seventeenth-century sailors unversed in the classics and unconcerned about matching their imaginary objects with actual star patterns. Drifting beneath unfamiliar skies, they simply named parcels of celestial real estate after the instruments that brought them there and whatever new birds and animals they came across. That's why we're saddled with essentially vacant parcels of southern sky designated Sextans the Sextant and Toucana the Toucan. The ancient mariners could not have known that their obsession with gadgets would drive future sky watchers daft in vain attempts to force a blank region named Norma to look like the Carpenter's Square or identify Antlia the Air Pump.

But we're stuck with this starry warehouse of antique contraptions. Linking the faint southern stars into new patterns is no solution. Switching to yet more Greek mythology to match the scheme of the northern sky would mean falling back on esoteric legends that make even less sense than the famous ones, where gods routinely meted out elaborate punishments (like chaining beautiful women to rocks) for minor offenses that should have been thrown out of traffic court.

Nor can we save the situation by substituting modern gadgets for the archaic ones. It's never wise to immortalize technology; not only do our artifacts fail to stand the test of time but enshrining them would suggest to future generations that we worship deities like Microsoft and Radio Shack. Anyway, would the constellations of Laptopium or Cellphonium really be an improvement?

When they do resemble anything at all, the constellations appear entirely different from what their names suggest. Cassiopeia doesn't look like a queen on a throne but rather the letter M or W, depending on the season. Virgo doesn't look like a virgin (I'm not going there) but instead resembles the letter Y. Sagittarius is so like a teapot that mapmakers have essentially given up trying to force an archer out of this pattern. On most modern star charts it *is* labeled a teapot, though no horoscope believer would own up to such a sign. ("You're a Gemini? Well, I'm a little teapot.")

Savvy skywatchers don't even attempt to envision such formerly popular college majors as "Herdsman" or "Charioteer"; when they want to spot the latter (Auriga) they look for a pentagon instead. Nor is it still fashionable to pretend to see a Big Bear in the stars of Ursa Major: It's the Dipper. Or if you live in England, the Plow. (Or Plough.)

Here's a list of discernible shapes in place of the obscure classical patterns.

OLD FIGURES	ACTUAL OBJECTS
Aries (Ram)	Three stars
Auriga (Charioteer)	Pentagon
Canes Venatici (Hunting Dogs)	Two stars
Capricornus (Sea Goat)	New York State
Cassiopeia (Queen on a Throne)	Letter M
Cepheus (King)	Another pentagon
Coma Berenices (Queen's Hair)	Buckshot
Delphinus (Dolphin)	Kite
Sagittarius (Archer)	Teapot
Taurus (Bull)	Letter V
Virgo (Virgin)	Letter Y

The best strategy is simply to form your own shapes and learn them that way. The "infallible" method of star learning, which works most of the time, involves grasping three simple sky regions and branching out from there. Three large regions are plenty, because as the sky rotates you're sure to have two in view at any season and any time of night. They will serve as splendid jumping-off points to more obscure constellations.

Zone One is found halfway up the northern sky, and it's always visible. It's the one you can dependably turn to, because it's the only one that barely shifts position.

You start with the Big Dipper: It's always in the north, high up in the spring and low in the fall. (Good mnemonic: Fall low and spring high.)

That so many disparate cultures—Greeks, Native Americans, Germanic tribes—all saw a bear here is astonishing.

Use the Dipper as home base. Follow the two end stars of its bowl to point downward (in the spring) or upward (in the fall) to Polaris, the North Star. Polaris is at the end of the Little Dipper's handle, a curved string of faint stars you'll find easily once you've escaped city light pollution by moving to Greenland.

In spring you can follow those leftmost Big Dipper pointer stars the wrong way, upward, to Leo the Lion; they go nowhere else. Next, look at the one medium-faint star between the Big Dipper's handle and the Little Dipper's bowl, and you've found Thuban, the famous pole star of 5,000 years ago. It's Draco the Dragon's most famous luminary. That skimpy string of faint stars weaving between the Dippers is all Dragonstuff.

When the Big Dipper is high, look low in the north for the "M" of Cassiopeia. Through spring and summer, follow the arc of the Big Dipper's handle to the brilliant orange star Arcturus (in the abstract constellation Boötes, supposedly a "herdsman" but actually more like a teenager's room). Continue that huge arc farther to the only low southerly brightish star, the blue Spica, the main luminary of Virgo. Suddenly you've got seven constellations. What did that take? Two minutes?

Zone Two: Orion's neighborhood.

In winter and early spring, the three-stars-in-a-row of Orion's belt are unmistakable. The bright orange star upper-left of the belt, marking Orion's shoulder, is the famous Betelgeuse. A scale model representing that supergiant as a sphere twenty stories high would make Earth the period at the end of this sentence. Using Orion's bright stars, it's easy to envision a hunter—or an environmentalist, a serial killer, or any person whatsoever. But the Babylonians saw a "sheep" here. Orion the Sheep—no, that just didn't cut it.

Follow the belt down to the left to Sirius, the Dog Star, and you've found Canis Major. This "Big Dog" was considered unlucky for centuries; Virgil wrote that Sirius "saddens the sky with inauspicious light," and Dante spoke of "the curse of days canicular." To Orion's upper left float two stars of almost equal brightness, the Gemini twins. They must be fraternal, because Pollux is brighter than Castor and slightly orange. Halfway between Pollux and dazzling Sirius sits the brilliant Procyon, which, when linked with the only nearby little star, produces Canis Minor, the minor dog. Two stars, and you're supposed to see a dog here? Better make it a Chihuahua.

Due north of Orion is winter's highest bright star, the creamy white Capella. Then follow Orion's belt rightward to the "V" of Taurus the Bull, dominated by the bright orange star Aldebaran. Next, everything below Orion's bright blue "foot" star, Rigel, is Lepus the Hare. A bunny? Sure. After two or three shots of tequila.

Presto: Seven more constellations. As easy as looking at your boss and thinking, "What an idiot!"

For Zone Three, we turn to summertime, which can be more confusing, because then the Milky Way tumbles across the sky, spilling a myriad of faint stars like fireflies. Look almost straight up in midsummer and there's the famous Vega (VEE-guh, not VAY-guh). This brilliant, blue-white star lies amid a quintet of fainter ones that comprise Lyra the Harp. Four of these form a perfect parallelogram; envision strings between them and you're seeing the sky's only musical instrument. (Vega was where the alien transmitter in the 1997 movie *Contact* was situated.)

The brightest star southeast of Vega is Altair, of Aquila the Eagle. Keep going a bit farther in that direction to the little kite-like quintet of Delphinus the Dolphin. Summer and fall's final bright high-up star is Deneb, the tail of Cygnus the Swan, east of Vega. You're now looking toward Cygnus X-1, the most famous black hole in the heavens.

Now look south from Vega to the only bright, lowish orange-red star of the warm season; you've found the heart of Scorpius, the great Antares. The Scorpion is probably the most realistic of all the constellations, but from most of the U.S. its tail hugs the horizon like a vacuum cleaner. Left of Scorpius is the teapot of Sagittarius, home to the center of the Milky Way. The "steam" pouring out of the spout is the very location of our galaxy's heart. Everything in the sky pivots around this place every 240 million years.

Boom! Six more constellations. And there you have it: the top twenty. Together they embrace every bright (first magnitude) star seen from the United States or Europe, except for autumn's solitary Fomalhaut, low in the south.

With just a little effort, you've captured the sky's elite constellations, joining the ancient desert-dwelling Arabs, to say nothing of Plato, Milton, Keats, and untold lovers who've sighed beneath the stars. You're one of them now—this generation's link in the unbroken chain of stargazers who have sought to know and revere the denizens of the night sky.

Sure, the patterns are strange—given human history, could it have been otherwise?

Chapter 27

Cosmic Name Calling

Names have power. Catchy ones stay with us, dull ones earn obscurity. Which is why Reginald Dwight decided to call himself Elton John and Cherilyn Sarkisian became Cher. But when it came to labeling the universe's inventory, phonetic ease and majestic nuance played no role. We established systems every bit as goofy and erratic as those we imposed on the constellations.

Take the stars. Only a few dozen of the ancient names are still in common use. Some of these sound dignified (Arcturus), some ridiculous (Zubeneschamali), some disconcerting to pronounce (Scheat). Some—for example, the Big Dipper pointer star Dubhe—are laughable when said aloud ("There's Dubby!"). Attempting to justify odd names by explaining their derivations (mostly from the Arabic), doesn't usually help. Tell people that "Betelgeuse" means "Armpit of the Sheep," "Deneb" means "chicken's tail" and "Fomalhaut" is "Mouth of the Fish," and the stars may still lack the glamour of the Hollywood variety.

At least such proper names provoke interest and recall ancient mythologies. The medium-bright stars, bearing Greek letters, lack that advantage. Still, they're logical and sound stately: No alien would be ashamed to come from Delta Leonis or Zeta Centauri. Then comes the third category, the faint stars comprising the vast majority, where competing numbering systems run amok. You can call the star orbiting the Cygnus black hole HDE226868 or choose BD+34 3815. You can even use designations from a host of disparate catalogs still widely used: The fifth and sixth nearest stars to Earth are commonly called Lalande 21185 and Wolf 359. Somehow such inconsistent terminology remains in place to this day. In any event, few of the blazing suns illuminating their celestial fiefdoms have titles worthy of their grandeur.

Move to the planets. Except for Earth, a word derived from Old High German, all were named for Roman gods, with Uranus alone shared by the Greeks. But their features seem to have been named by people who munched on too much peeling paint. The Red Spot. The B Ring. Snoopy. The Dark Spot. Who's responsible for this?

The International Astronomical Union, that's who. They alone are empowered to name the contents of the cosmos. Labels that existed before this organization's mandate were not changed. Assorted tiny features (like boulders near parked spacecraft) have been named by NASA staff or schoolchildren (hence the Yogi rock on Mars).

In recent years the IAU has set strict naming guidelines for all newly found solar system features, such as mountain chains, plateaus, and craters. Listing all the schemes would require pages, but consider just the rules for a few of Saturn's eighteen principal moons:

Titan's features are to be named for ancient displaced cultures. Iapetus's will glorify people and places from Dorothy L. Sayers's translation of *Chanson de Roland.* Rhea: people and places from creation myths. Dione: people and places from Virgil's *Aeneid.* Enceladus: people and places from Richard Burton's *Arabian Nights.* Mimas: people and places from Malory's *Le Morte D'Arthur* (Keith Baines's translation). And on it goes. No movie stars, no contemporary authors, no cartoon characters. In fact, nothing whimsical at all. Touring the solar system will be like trying to stay awake during freshman English.

Many moons, however, were named before the IAU arrived on the scene and they should please anyone who finds inconsistency delightful. Mars's satellites Fear and Death (Phobos and Deimos) take the "Most Depressing" prize and come from Greek mythology, as do most planetary moons. But Uranus's satellites are all characters from Shakespearean plays, which is why there's an actual place in space named Puck.

But what about our Moon? The blotches are all maria, a permanent legacy of the many centuries when people thought they were oceans. Features on the lunar far side all have Russian names, because that country's *Luna 3* arrived there first, in 1959. Full moons themselves have names, but there's no standardization. Newscasters sometimes authoritatively announce that a particular upcoming full moon is called the "Wolf Moon," or the "Strawberry Moon," or whatever. In reality, cultures around the world have had many names for the thirteen full moons of the calendar year. Most in the United States and Canada come from Native American tribes, especially the Algonquin. Few such names reflect the full moon's appearance, motion, or behavior; instead they describe natural seasonal conditions or activities associated with that time of year. January's full moon was called the Winter Moon or Yule Moon by the colonists, the Wolf Moon or Old Moon by the Algonquin, the Moon of Frost in the Tepee by the Lakota Sioux, the Ice Moon by the San Juan, the Cold Weather Moon by the Nez Perce, the Hoop and Stick Game Moon by the Cheyenne, and so on. Be wary of media announcements claiming that the current full moon has a particular official name. Only the Harvest Moon (the full moon nearest to the autumn equinox) and the Hunters Moon (the full moon after that) are established labels in widespread use.

The "Blue Moon" has only recently come into the language. It first appeared in the 1930s in an astronomy magazine as a falsely attributed citation, and was subsequently repeated often enough to establish its definition as the second full moon in a given month. Because the lunar phase cycle of 29.5 days is shorter than all calendar months except February, a second full moon usually occurs whenever a full moon falls on a month's first or second day. As this happens every two and a half years on average, a "blue moon" is not particularly rare. An actual visual blue moon caused by chemical air pollution fortunately occurs much less often.

Asteroids are another story. First named for figures in Greek mythology, they recently became contestants in a modern Miss Universe contest judged by the IAU's naming committee. Favorite deceased astronomers and physicists are honored with their own chunks of rock. Shoemaker floats next to Ceres and Eros. Speaking of which, Eros is one of only a handful of heavenly bodies that the IAU loosened up for. Craters there shall be named for "mythological and legendary names of an erotic nature." How did *that* category make it through? Must have been a late-night session. In any event, nobody will have to guess where to bestow the name Lewinsky.

Meteor showers are named for constellations (the Perseids, the Leonids) and meteorites for whatever place on Earth they smashed into (ALH84001, of the controversial Martian fossils cited in chapter 12, was labeled for the Allen Hills region of Antarctica). Comets are the sole objects named for the person who first reported them. This is the only legitimate way to get your name emblazoned in the heavens. If the comet brightens spectacularly, your name will be famous everywhere, and you become the sole authority on how it should be pronounced.

(Of course, the sky is not unique in being saddled with mind-numbing designations. Earthly life-forms don't fare much better. Biologists and zoologists have two systems to cope with. That's not merely a Norway maple tree, it's also an *Acer purpureum.* You may spot a rat, but to a zoologist it's *Rattus rattus.* A few years ago the official world body that decrees such things decided that rabbits should no longer be lumped in the rodent category, so now they have their own order. Rabbits are lagomorphs. Big hopping *whoop!*)

Arguably the loveliest big things in the cosmos are the floating Rorschach tests called nebulae. At some point, some nameless astronomer imagined that a particular gas cloud looked like gym equipment, and the "Dumbbell Nebula" label stuck. As did the appellations of eighteenth- and nineteenth-century astronomers who saw an Eskimo and a Crab. Apparently much observing was done before the "Just Say No" campaign.

Though more backyard telescopes are now routinely trained on the heavens than ever before, the Name That Nebula competition seems to have lost steam. There are still lots of nameless blobs of dusty hydrogen out there, but the status quo is now fixed. Too bad. It would be nice to spice the firmament with trendy labels (the Matrix Nebula), confer honorifics (the Captain Picard Nebula), or tell future generations about important aspects of our lives (the Burger Nebula).

Saved for last are the largest and grandest structures in the cosmos— galaxies. Each type of celestial object has its own separate naming system, but in this menagerie galaxies are unique, because until the 1920s only a single example was known: ours. Imbedded within it, unaware of its true nature, many cultures regarded the glowing sky belt as a stream of spilled cream, so it was given the Latin name Via Galactica: Milk Road, or Milky Way. There was only one, and it was thought to constitute the entire universe.

In the 1920s, when other such realms were finally recognized, they were called "galaxies," retaining the milky theme, as though the whole universe were one vast dairyland. (There's "lactic" in "galactic.")

But how to name them? Since consistency within categories is normally enforced, we could have kept the Milky Way motif and given new galaxies labels like Twix and Almond Joy. Or we might have milked that dairy theme by naming them after various cheeses, or ice cream flavors—which would have involved locating supernovae in Butter Pecan 432 or Pistachio 6695. Instead, only a few dozen galaxies wound up with any sort of name at all, and uniformity played no role. Some were named for articles of clothing (the Sombrero Galaxy) or injuries (the Black Eye Galaxy). The huge majority merely received number designations, like NGC4565 (from the New General Catalog). Which makes sense, since there are 200 billion galaxies in the observable universe, while English has only 200,000 words and most are poor potential candidates, like "however" and "weasel." There's no way all known galaxies could be named, unless each and every English word was shared by a million galaxies, as in Weasel334792, but then we'd be back to mostly numbers anyway.

So the universe's inventory will stay just the way it is. We took no part in its curious, inconsistent, and sometimes comical naming, but that's OK. It will survive.

Stars for Sale

It's a common experience. You're absentmindedly listening to the radio when something snaps you to attention. Around the holidays it may well be the announcer saying, "If you're looking for a perfect gift for that special someone . . . have a star named for them!"

You might wonder how the International Star Registry, the advertiser, remains in business. Twenty-five years ago, serious sky enthusiasts thought Ira Downings of Toronto, the person behind it, would make a quick buck and then vanish. Wrong. After he sold the franchise to U.S. entrepreneur Phyllis Mosele, the now Illinois-based outfit grew bigger than ever and began running radio commercials. They're on the Web, too. A pioneer in the cosmic name-peddling market, ISR has labeled over a half million stars since the enterprise began in 1979. The business is so lucrative that their lawyers have to fend off other star-naming wannabes, giving new meaning to name-calling fights.

Valentine's Day and the Christmas holidays are their peak season, according to general manager Mary Ellen Mottyka, and no surprise: The gift is both impressive and romantic. It lets you turn to the person you love, sweep a hand toward the heavens and declare, "Up there, sweetheart, is a star that now bears your name forever." But you can't gesture too precisely: The star you've bought is dimmer than 11th magnitude and hard to find even with a small telescope. ISR ran out of naked-eye stars years ago. (There are far fewer of those than most people imagine. Even in a clear rural sky only 6,000 stars can be seen without optical aid, or roughly 2,500 at any given time. The heavens may seem crowded with "millions," but you could count every star, even the faint ones that seem sprinkled everywhere, in less than half an hour!) Still, now that you know where the constellations are, at least you can point in the general direction.

Romantics and the naive may buy into it, but the New York City Department of Consumer Affairs is less starry-eyed. In 1998 they issued a violation notice to the International Star Registry citing a "deceptive trade practice." You're not supposed to sell something you don't own, and the ISR is not an astronomical body of any kind, let alone one with any connection to the IAU. To that dignified organization, name-selling is about as appealing as renaming the planets after *Simpsons* characters.

Of course, the average person doesn't know this, and customers continually respond to ISR ads, which reek with authenticity. The company promises that your name will be copyrighted, protected in a vault, and that the star bearing your name is "located using information that guides the Hubble Space Telescope." For sending $48 "plus shipping and handling," purchasers receive a one-page star chart with a single dot circled. They also get a fancy parchment proclaiming that henceforth this star will be known as Brittany or Taylor. Not too long ago the certificates also claimed inclusion in the Library of Congress, until the company was prohibited from saying that. Even so, the enterprise seems as eternal as the firmament itself.

In truth, one little thing is missing, as the New York consumer watchdog agency noted. No astronomer, no observatory, no actual star atlas, nobody remotely connected with astronomy will ever call this star "Brittany," even if it explodes into a supernova that lights up the sky. Every astronomer and all the media will still refer to it as HDE337962. Who, then, will identify the star with your name? Only the ISR.

This would seem a stopper, but no. In fact, the ISR can stay in business because there really is no law against renaming anything as long as you avoid trademarks or exploiting celebrity names. Indeed, nothing stops you or me from boldly taking the idea a giant step further and offering to bestow someone's name on an entire galaxy. And why not? There are billions of known galaxies and most have no designation whatsoever. Anyone is allowed to call anything anything. So, while it's still legal, let me offer you a much better deal than the International Star Registry. Send in a mere twenty bucks and my observatory will allude to *the Sun* with your name. (Once, that is.) Then on the appointed morning you can throw open the curtains and announce "Tom has risen!" Why settle for an 11th-magnitude star when your name can be attached to an object that you can not only see but that *helps* you see?

Entrepreneurial opportunities seem as limitless as the expanding universe. Why isn't anyone selling the celestial bodies themselves instead of merely their names? (One Christmas long ago, my parents gave me a framed official title to an acre of land on the Moon, a popular gift at the time. I treasured that certificate for years, until I finally realized, with a slow-wittedness my friends assure me I still exhibit, that it was just a gag. And that's why I can probably get away with attempting to sell the Sun: Everyone knows I don't own it. Even if you did buy it, where would you keep it?) The ISR irks the science community precisely because the company's official-sounding name and the insinuation of legitimacy in its ads implies that the new star name will be accepted and used by astronomers. And that just ain't so.

In an era when almost everything has a price, it's reassuring to know that the IAU remains opposed to jumping on this lucrative bandwagon. Yet think of the research that could be funded if the world's only legitimate star-naming body sold off its star names. The current official practice of bestowing stupefying letter-and-number designations on stars could hardly be harmed by calling a few of them Jennifer or Courtney. We could use some fresh names out there. How unimaginative that the magnificent rings of Saturn are named A, B, C, and so on, and that Jupiter's ancient, violent, Earth-dwarfing cyclonic storm is called "The Red Spot." There's always been a yawning gap between the epic, lunatic architecture of the universe and the dreary logic of our designations. Let's nominate a few poets to sit on the IAU's naming committee.

Meanwhile, at this moment, as many as 2,000 amateur astronomers in countries where night has fallen are staring into telescope eyepieces. They hope to get their name attached to the comet they are intensely attempting to find in the vastness of the heavens. The competition has endured non-stop for over four centuries and produces only a handful of winners per lifetime. No wonder so many others just drop a $48 check in the mail.

Chapter 28

BANG!!!

The great mystery of cosmic birth has tantalized human minds since the brain first harbored thought. What was the primordial urn that held all the matter in the universe? What triggered its emergence? Remorseless cycles of speculation have currently pushed metaphysical musings out of popular fashion, so, leaping over that particular patch of quicksand, let us deal solely with what is scientifically known or suspected.

In the public mind, the Steady State theory still somewhat competes with the Big Bang as an explanation of the universe's genesis. Older than science itself, the notion of an eternally existing "steady" universe is widely associated with Sir Fred Hoyle, who until his death in 2001 crusaded for its acceptance. But to cosmologists the issue has been resolved since the 1960s. Despite significant nagging problems, the Big Bang has become the most credible explanation for the beginning, some 14 billion years ago, of all the strangeness transpiring thereafter.

It requires a big stretch to dismiss the Big Bang. Not that a "steady state" is inherently unworkable. If just a single hydrogen atom popped up each decade in each segment of space with the volume of a football stadium, this could supply the material needed to fill the gaps left by an expanding universe. Since there is no way that the appearance of such a minuscule bit of extra material could be detected (that volume of space already contains over a trillion hydrogen atoms on average; one more would hardly set off alarm bells), how might the idea of this kind of steady-state universe be disproved?

The springing-out-of-nothingness of one atom at a time isn't a particularly vexing issue in itself. When it comes right down to it, no one ought to care whether the universe leaks in—matter trickling into existence atom by atom, like water dripping from a faucet with a worn washer—or bursts full blown onto the scene. Neither scheme makes sense. Happily, there's an easy test of the opposing slow-but-steady versus *whammo* theories. If the latter is true, then the universe's birth occurred exclusively in the past, and the past should therefore look different from the present.

As we know, viewing objects in the distance is tantamount to viewing the past, since it takes time for images to reach us. What we find is that distant galaxies do appear different (they're bluer and more of them are spiral) from those nearer to us; a cosmic evolution is apparent. Moreover, researchers using the Hubble Space Telescope and the Chandra X-ray instrument in 2003 found ample evidence of a universe-wide evolution. Galaxies across the cosmos steadily increased in size during the first 7 billion years of the universe's life, and star formation was extremely rapid. Also during that period the universe decreased its rate of expansion; gravity prevailed, slowing down the universe's growth. Then, some 6 billion to 7 billion years ago, when the universe was half its present age, everything changed. Star formation quickly dropped to just a tenth of its former rate, while the universe reversed its previous braking and started to increase its dimensions at an ever accelerating pace. (The change-over date to rapid expansion is controversial; some astrophysicists think it started only very recently, and that the slowing went on for almost all of the universe's past 13.7 billion years.) It was as if the force of gravity had suddenly been overcome by an outward-pushing force, a sort of antigravity. An epic development, and a weird one.

The Big Bang also furnishes an explanation for the ratio of elements seen in the universe; it predicts that in the fantastically hot and dense environment that prevailed in the first few minutes of the present universe (mostly when it was between 100 and 1,000 seconds old), hydrogen, helium, and lithium formed in just the proportions seen today. Heavier elements than these required star cores for their birth, but the three lightest elements would have had to "be there" almost from the get-go, even before stars were born. The prevailing 3-to-1 ratio of hydrogen to helium is right on target if the Big Bang happened.

Then there's the cosmic background radiation, discovered by Robert Wilson and Arno Penzias in 1965 while using a Bell Laboratories radio telescope for an entirely different purpose. (Who would have imagined that the birth of the universe would have been discovered in New Jersey?) Surveying the sky's radio background in preparation for planned communications satellites, Wilson and Penzias found omnipresent noise they could not explain. At first they assumed it was caused by heat from droppings left behind by the many pigeons roosting in the metallic horn-shaped antenna, but repeated cleanings didn't clear up the static. Their colleague Ivan Kaminow later remarked to a reporter, "They looked for dung and found gold, which is just the opposite of the experience of most of us."

In 1927, the Belgian astronomer George Lemaître was the first to declare that the universe started as a small, hot, superdense ball. He called it a "cosmic egg." Two decades later, physicist George Gamow and others predicted that if a Big Bang really occurred, it should reveal itself with leftover background heat that, because of the universe's expansion, should now have been redshifted and thus weakened. The detection of just such a uniform radiation, albeit quite a bit chillier than initial predictions, provided the Big Bang with additional dramatic support.

Eventually, finer measurements of the radiation, first made with the COBE (Cosmic Background Explorer) satellite in 1992, and refined with the Wilkinson MAP (Microwave Anisotropy Probe) launched in 2001, showed that the 2.73° Kelvin (or -454°F) microwave background temperature displays a remarkable uniformity in all directions around the sky. Tiny anisotropies, or ripples, in this radiation tell epic, eons-old stories about the universe's infancy. (In this case, "tiny" amounts to deviations from uniformity of 1 part in 100,000.) These minor irregularities are widely regarded as seeds of incipient clumping from which eventually sprang galaxies and the other structures of the universe.

An important 1979 refinement posited a universe whose space wildly inflated during the first fraction of a second, imparting a uniformity that would otherwise have been impossible. Other evidence agrees with the theory that 380,000 years after the Bang the hot, dense, primeval fog cleared; space suddenly became transparent to light. The earliest detectable energies, the cosmic microwave background, date from that moment, which was the true "Let there be light" episode.

Such clues, plus the simple fact that galaxy clusters are all racing away from one another, makes the Big Bang theory a tough act to follow.

And though we've gotten a lot of bucks from the Bang, the cosmic genesis jigsaw has important missing pieces: Our understanding is still far from complete. Everything probably did begin with a bang, even if we whimper about the details well into the coming century.

Escape Clause

Although astronomers generally regard the Steady State theory as a beautiful idea that is as dead as the passenger pigeon, it certainly carries far greater esthetic appeal than its rival. A handful of cosmologists still search for a way to revive it.

Who could love the Big Bang (a name Hoyle himself coined disdainfully) when it promises only a cosmos that will either stop expanding, reverse direction, and eventually crush itself down to a size smaller than a pea, or else—much more likely—blow itself into isolated icy islands of frozen death? Neither sounds particularly attractive.

Tough luck, say physicists. The universe will perform its drama on its own terms; our sensibilities are irrelevant. Our trust in science has most of us pretty much resigned to either of the Bang's dreary denouements, consoled perhaps by the fact that humans will no longer be seated in the audience.

Yet many harbor a hopeful feeling that the same cosmos that conjured up Beethoven and basset hounds will not utterly self-destruct. Optimists argue that any sort of permanent turn-off-the-lights oblivion seems out of character with the universe's life-directed leitmotif.

Wishful thinking? Perhaps. But while the Big Bang looks like an established fact, there exists an escape clause that doesn't resort to ideas unworthy of science or logic. It requires only that we entertain the possibility that time can gradually change its rate of flow.

This is hardly a new notion, nor was its chief advocate any sort of crank. No less a physicist than Paul Dirac—who first predicted antimatter in 1928 and later won the Nobel prize—argued consistently that there's no reason that time should flow at a steady rate forever.

Time is strange to begin with, as will be seen in the next chapter. On many levels—and perhaps even on the most fundamental plane—it may not exist at all. But let's not make this unnecessarily mysterious. Rather than invoke other dimensions, we can simply deal with time's everyday guise as a sort of synchronicity between events. While half of a uranium sample decays, Earth performs 4.5 billion orbits of the Sun and Saturn circles our sky 153 million times. Those three events and countless others will continue to occur in sync. Dirac's question was: What if the whole shebang gradually slows down in unison and on every level? We wouldn't notice a thing, but it would have enormous cosmological implications.

If time ran differently long ago, we would still observe a cosmic background radiation that pointed to a Big Bang that would seem to have happened around 14 billion years ago. In reality, however (said Dirac), the universe would be infinitely old. There would not be an instant of creation; rather, the scheme would all revert to some variation of Hoyle's Steady State theory.

Variable time seems more likely now than it did a few years ago. In 2001, physicists uncovered evidence that the universe's physical constants can or do indeed change. Specifically, the strength of the electromagnetic force was found to have been smaller by a tiny fraction 8 billion years ago. Astrophysicists reached this conclusion after observing photons interact with distant nebulae on their way to Earth. Indeed, the value for the Planck constant (the consistent multiplier for energy leaving any radiating particle), which in turn derives its stability from the supposedly unchanging nature of three other constants, seems to have been slightly different back then. That astronomers measured this alteration to be exactly the same in many separate places gives the discovery far more credibility than if the effect had derived from a single observation.

If some constants can change over the eons, then why not time itself? Dirac was particularly suspicious of the stability of the gravitational constant, concluding that its force might be as undependable as a child's alibi. (If Dirac is right, the gentler gravity of the future might spare us a scraped knee when falling down a flight of stairs.) In fact, if time alone has changed, then other constants can seem to alter but actually be constant.

Tricky business. Even if it were happening, we would not expect to see any difference in the flow of time when we observe far-off galaxies. Their images have long been traveling to us, and during that vast en-route interval their incoming photons have also been caught up in the new rate of time passage. When we finally observe images of distant star-meadows, frames of that movie run through the projector of today's temporal pace, so that it appears that no time mutations have happened at all. How, then, can we know one way or the other?

Evidence on Earth might reveal it, except that our measurements of the way time passed here a million years ago are not precise enough to detect infinitesimal alterations from the way it flows today. Dirac himself could offer no decisive way to determine whether or not time is stable.

What do we do with all of this? Given credible evidence that one or more of the universe's constants may not be constant, but with no hint that time could be one of them, any present conclusions are limited. Still, those of us who teach science may well wonder whether we should present the Big Bang as a fait accompli.

It's probably safest to tell students that up to now evidence points strongly to a Big Bang that happened 13.7 billion years ago and, further, that the cosmos *appears* to be increasing its rate of expansion.

Refinements in distance-measuring have made researchers and theorists fairly certain that the cosmos is unlikely ever to come back together into a crunch, even one that offers rebound possibilities. That is, we don't currently seem to be mailing our letters in an oscillating universe that eternally inflates and deflates like the lungs of some cosmic Being, as in the old Hindu legend of the breaths of Brahma. (Besides the size of the cosmos, the other vital parameter necessary for determining future expansion possibilities is its total mass. The best present determination for the weight of the universe in tons appears to be a 1 followed by 50 zeros.)

Still, the door remains slightly ajar, and the prospect persists that the universe is infinitely larger than we imagine and never had a birth at all. It's too early to know for sure whether or not the game is played according to Hoyle.

We've only begun to decipher the rulebook.

Chapter 29

Travels in Spacetime

Time is often called the fourth dimension. This usually throws people, because in daily life time bears no resemblance to the three spatial realms, which (to review basic geometry) are:

Lines, which are one-dimensional.

Squares, circles, triangles, etc., which are two-dimensional because they enclose an area.

Spheres, cubes, and all their cousins, which contain volume and are therefore three-dimensional. But an actual cube requires four dimensions, because it persists and sometimes even changes. Thus something else besides the spatial coordinates is part of its existence, and we call this something "time."

Space and time do not exist independently of one another. Spheres and cubes and you and I travel through a combination of the two. Einstein was the first to realize that if this were not so, curious time and space contradictions would arise. Because of light's finite speed, an event like a couple of flash pictures taken on a fast-moving jet might seem to an outside observer to have occurred in reverse order. The only way to ensure that you, me, and everyone agrees about when events occur is to specify them in the matrix of spacetime.

Actually it was not Einstein who coined the term "spacetime": he didn't even like the concept at first. But his 1905 special relativity theory did rob space and time of their status as absolutes. It took three more years and

German theoretician Hermann Minkowski to take Einstein's theory and run with it, laying the foundation of an important refinement: that the cosmos is really woven of a four-dimensional spacetime.

How does this affect your daily comings and goings? Not one bit, for the simple reason that your own speed is very similar to that of all others around you (except perhaps for a few people you knew in high school). Any discrepancy in the dimensions of physical objects or in the time reported by you or others will be too small to notice.

Still, the discrepancies exist. We'll see in chapter 31 that a continuous rain of fast-decaying subatomic particles called muons shower down from the upper atmosphere into our bodies. We observe their time to be retarded compared with ours, allowing them to survive long enough to reach us. They would perceive (if they could perceive things) that time passes normally but that Earth's atmosphere is less than a half-mile thick. We disagree about time and space. But we'd agree about it if the muons' paths were plotted in spacetime.

As noted in chapter 4, according to Einstein's general theory of relativity (published in 1915) objects that move freely—that is, everything that falls, or is dropped, thrown, or otherwise not continuously propelled by some outside force—will simply travel in the shortest possible path through spacetime, a trajectory called a geodesic. Einstein realized that spacetime itself must warp if light is to preserve its constancy and if contradictions and paradoxes are to be avoided, and if all the laws of physics are to function everywhere in exactly the same manner without there existing any privileged position.

This concept alters how we visualize motion. Instead of picturing Earth orbiting the Sun because the Sun's gravity is pulling on us (and changing our track from a straight line), we can ignore gravity as a force altogether. Rather, anything with any weight at all will bend the spacetime in its vicinity. The more massive and compact the object, the more its surrounding spacetime will be curved. The Sun, which is the most massive object in our celestial vicinity, weighs some 333,000 times more than Earth, or about 2,000,000,000,000,000,000,000,000,000 tons. This is enough to bend the surrounding space to such a degree that Earth, falling in a straight line through this curved space, returns to its starting point after a year.

But such motions are not limited to space alone; so tightly are space and time bound in the formula that we must instead conceive of spacetime as a unique and cohesive entity—just as sodium and chlorine combine to make salt, something altogether dissimilar from either element by itself.

This linkage is an intimate one. If you sit in a chair, you do not travel through any space relative to the room around you, but you still pass through time. You must. You age, and you experience one day every twenty-four hours. If you get up and walk across the room, you begin to traverse spatial dimensions while continuing your motion through time, which now unfolds at an infinitesimally slower rate. If you could travel at the speed of light, you'd move through lots of space but would not travel through time at all. That's why, if you aim a flashlight out a window and turn it on, from each photon's perspective it has already reached the farthest edges of the universe. It has gotten there in no time. At the speed of light, time does not pass at all. The more we zoom through one component of spacetime, the less we move through the other.

Of course, we can still view space and time separately when the situation demands it, as when we perform everyday actions on our slow-moving planet. Nonetheless, our time passes a bit differently from the way it does on any celestial body with a different gravity. Each month, earthly clocks would lag behind timepieces on the moon, and the *Apollo* astronauts aged just a little faster than their relatives back at home.

In very strong gravitational fields, spacetime itself has a warped, non-Euclidean geometry. Parallel lines do not meet, and the sum of all angles of a triangle do not add up to 180°. It's as if the triangles were inscribed on a curved surface, even though no surface is visible.

General relativity continues to be tested whenever opportunities for finer measurements present themselves. One occurred in the mid 1970s, when the two *Viking* spacecraft were orbiting Mars. As the Red Planet slid behind the Sun, radio signals received from the craft were suddenly delayed by several microseconds, as if Mars and the spacecraft had abruptly jumped eighteen miles farther away from us! This result, caused by the longer path taken through the more intensely curved space near the solar limb, agreed with the predictions of relativity to 1 part in 1,000.

But absent any matter or energy to define its shenanigans, curved space is meaningless. Perhaps it's best to see it as a symbiotic relationship: Mass tells spacetime how to curve, and curved spacetime dictates how the mass must behave.

It's intriguing to picture even ordinary "flat" spacetime in operation. Take, for example, two batters who follow each other to the plate. The first may hit a line drive to the second baseman, while the second hits a pop-up that soars high and stays aloft for many seconds before coming down, again into the busy second baseman's glove. In our everyday perception of space and time as separate entities, these seem like different types of occurrence. But general relativity states that anytime an object is released to travel on its own, it must follow the same geodesic, or path, through spacetime as any other object in the same gravitational field. Translation: These two events— a fast line drive and a slow pop-up—take identical paths through spacetime. They march along geodesics of the same shape. By graphing both space and time as aspects of a single continuum, we see that the pop-up occupied a longer path through space but took more time to do so, while the line drive was straighter but consumed less time. When combined, the spacetime arcs produced by each batter have exactly the same profile.

But Is It Real?

Is time an actuality in and of itself—or just an idea, a way of expressing motion mathematically? The question is much trickier than it may seem.

In physics, time appears to be indispensable in just one area—thermo-dynamics, whose second law has no meaning at all without the passage of time. The second law of thermodynamics describes entropy, the process of going from greater to lesser order (like your desk drawer). Without time, entropy doesn't happen or even make sense.

Consider a room containing pure oxygen, with nitrogen filling an adjacent room. Open the door and they'll mix. Even if you waited forever, the original ordering of molecules would never restore itself. This evolution away from structure and toward sameness or randomness is entropy. The process pervades the universe. According to most physicists, it will prevail in the long run. Today we see individual hot spots—stars—leaking heat and spewing subatomic particles into their frigid environs. The organization and structure that now exists is slowly dissolving, and this entropy, this overall loss of structure, is on the largest scale a one-way process.

Entropy suggests a directionality of time, because it is a nonreversible mechanism. In fact, entropy defines the arrow of time. Without entropy, time need not exist at all.

And maybe it doesn't exist anyway, say those who argue that time on the deepest levels of reality has no validity or purpose (except, whimsically, to keep everything from happening at once). Newton's laws, relativity, quantum mechanics—all function independently of time; that is, they operate backward as easily as forward.

Metaphysicians, taking entirely different routes, have also questioned time's validity. The past, they say, is just an idea in a person's mind; it is no more than a collection of thoughts, each of which occurs in the present moment. The future is similarly nothing more than a mental construct, an anticipation. Thinking is a neuroelectrical process that occurs strictly in the here-and-now. So if our minds' ability to link experiences with memory or anticipation is independent of time, does time then exist anyway?

I'll refrain from entering this familiar debate. Suffice it to repeat: Thermodynamics provides the primary argument for time's reality—although some physicists discount the entropy argument, arguing that it is only we humans who define what is order and what isn't. Far more states exist in what we would term "disorder." Entropy, according to this way of thinking, is just ordinary random behavior and therefore not time-dependent.

In 1998, physicists at CERN in Geneva found a small asymmetry in the way particles and their antimatter analogs (kaons and anti-kaons) transform into each other, and this appears to provide evidence for time's existence. (Confirmed later that year in the US at Fermilab, it may also solve one of cosmology's greatest puzzles—why the universe is mostly matter when the Big Bang should have created equal quantities of matter and antimatter.)

The answer may be migraine-fueling in its complexity. For all we know, there may be many planes of physical reality, and time may operate on some levels yet be nonexistent or irrelevant on others.

Physicists in the past two or three decades have also taken seriously the notion that the "arrow of time" can change direction. Even Stephen Hawking once believed that if and when the universe started to contract, time would run backward. But he later changed his mind, as if to demonstrate the process. In any event, time running backward is not as daft as it first seems.

We protest because we think it means that effects will precede their causes, which can never make sense. A car accident would become a macabre affair in which injured people instantly healed without a blemish while their wrecked vehicle uncrumpled and repaired itself seamlessly. This is not only ridiculous, it doesn't accomplish any purpose, such as, in this instance, instruction in the evils of using cell phones while driving. The usual answer to such objections is that if time ran backward, our own mental processes would operate in reverse as well, so we'd never notice anything amiss. So they say.

Assuming that time is an actual state of existence, might it not follow that time travel should be valid as well? No! Theoreticians seem unanimous in dismissing the possibility of time travel, or even of other temporal dimensions existing in parallel with ours. Aside from the violations of known physical laws, there's this little detail: If time travel will ever be possible, so that people can journey into the past, then—where are they? We've never been faced with tales of unexplained people arriving from the future.

The Right Time and Place

The price has fallen to where it's affordable by almost everyone. The bargain we're talking about is GPS, the best thing to come down the pike since the TV mute button.

You already know, of course, that a handheld Global Positioning System unit picks up signals from space satellites to calculate your precise location. In 1999, the U.S. military, which operates the twenty-four satellites comprising the system, turned off the deliberate degrading or blurring designed to prevent enemies from programming missiles to hit targets in the United States. Instantly, everyone's GPS enjoyed a fivefold accuracy boost, with a resolution of ten yards, or two yards using special techniques.

Two paces! Your handheld unit receives signals from 11,000 miles up and then yields a position that has a 90-percent probability of being correct to within the length of a bed!

Suddenly, all other navigation instruments on an aircraft's panel have been rendered obsolete except as a backup. Fogbound mariners groping around on the bay don't have to wonder where or how far their marina lies. Hikers won't get lost, or need to wonder which mountaintop they've just reached.

Cars increasingly come with GPS installed. The deblurring has upped the resolution to where it not only shows you that you're on the freeway but that you're in a northbound lane. It tells you exactly which off-ramp you're approaching, and if you blow the exit it will show you the best way to make up for your mistake.

And it's all free, without monthly fees for usage—at least for now. But how it works is just as extraordinary as what it does.

The system's two dozen satellites are not geostationary but move slowly in twelve-hour orbits designed to keep at least five satellites constantly above a user's horizon. Each sends a time signal tuned to atomic clocks—ultra-accurate time that is continually monitored, and corrected as the need arises. Your own unit receives ticktocks from three or more satellites, and then invariably discovers that the time from each satellite is . . . wrong!

It's wrong because the signal, despite racing at the speed of light, requires a billionth of a second to travel each foot of distance to you. A satellite straight overhead is closest and thus delivers a signal closest to "true" time, while satellites on the horizon give you "older" time. Your unit picks up several, uses the fact that each nanosecond deviation from "true time" means that the satellite is an additional foot away from you, and thus calculates precise distances to several satellites. Then it quickly solves this mathematical problem: What is the only place on Earth where you would be that specific distance from each? (There are actually two possible solutions, but the unit disregards the one that would place you thousands of miles from where you were the last time you used it.)

Repeating the calculations every second, a GPS unit shows the heading and speed you're walking and even correctly revises its readout within a second or two of your changing direction. Just imagine: If you stop to pat your pockets to see if you've forgotten your keys, satellites far out in space can tell in a second that you've momentarily discontinued your stroll.

And it's all even more technically astounding than that. Because of six separate effects predicted by relativity theory, time itself (as experienced by the satellites and/or by your unit) runs at contradictory rates that would render the system useless if it weren't continually being corrected. For example, you might use a GPS in Alaska, where Earth spins slowly, or in the tropics, where your body whirls at over 1,000 miles an hour: Speed changes the flow of time, as we've seen. Other Einsteinian parameters also necessitate continuous and precise compensation. So the next time you imagine that relativity is just a theory, consider GPS. It would be useless if the slow-down of time caused by stronger gravity and higher speed were not factored into the system's routine operations. When it comes to global positioning, we'd be lost without Einstein.

Atomic Time

Atomic time, vital for the GPS system, has also become a fixture in many homes. For under $30, anyone can buy wall clocks (even wristwatches) that receive time signals from the master atomic clock at the National Institute of Standards and Technology, in Boulder, Colorado. Even accounting for transmission delays, the signal is true to within a hundredth of a second. Accurate enough to know when your favorite TV show is on.

How do the clock watchers at NIST know they've got the right time? First, they depend on a European agency that tracks our planet's spin. Called, not surprisingly, the Earth Rotation Service, it averages signals from 150 atomic clocks around the world and uses that to pronounce the official time, termed UTC (Universal Time Conversion). Then NIST broadcasts it to the rest of us.

NIST itself operates a bevy of cesium atomic clocks, including their newest pride and joy, NIST-F1. It will be 20 million years before it gains or loses a single second, but NIST will never let it get that far off. A second to the fellows at NIST is like a week to us. (Donald Sullivan, who runs NIST's Time and Frequency Division, nonetheless confessed to me that he keeps his watch a few minutes fast to be on time for meetings.)

Cesium atomic clocks work by keeping a series of gaseous cesium atoms suspended in a chamber and monitoring the vibrations of each atom's nucleus. The cesium nucleus has two possible ways of spinning, and will maintain the more unlikely state only if it is spoon-fed energy with the same number of vibrations as the atom's own natural resonance rate. That figure: 9,192,631,770 per second. (Of course you'll remember that.) A second is *defined* as that many vibrations of the cesium-133 atom. In effect, each official second is divided into that many ticks.

An infrared signal with exactly this frequency constantly bathes the ultracooled cesium atoms. If the frequency wanders, the cesium flips over to the other type of spin, which sounds an alarm to tweak the frequency back to where the cesium spin stays put. The entire apparatus is set up to adjust itself automatically.

The better cesium clocks have fancy ways of suspending each bit of cesium while it's being bathed in the infrared frequency. But if you're not fussy and want an older cesium atomic clock (like the NBS-1, circa 1951, a Wurlitzer-type antique that is accurate only to 1/100,000th of a second a day) you can pick one up for $50,000 or so at one of NIST's yard sales. Don't expect a compact night-table model; atomic clocks are about the size of large refrigerators.

Best bet is just to buy a clock that picks up NIST's signals. It will even set itself to Daylight Saving Time and take care of leap seconds. In the end, it's a time saver.

Leap Seconds

Most years, an extra second is tacked onto December's final minute. As of January 1, 2003, twenty-two have been inserted since the process began in 1972. You don't see this during those televised New Year's celebrations, but the official clocks really read 23:59:58, 23:59:59, 23:59:60, 00:00:00. And in the kickoff year of 1972 that extra second was inserted twice, making 1972 the longest year in history.

Newspapers periodically take it upon themselves to try to explain why leap seconds are necessary. No one seems to get it right, either in print or broadcast. They tell us that leap seconds are needed because our planet's rotation is slowing down a second per year. Not so: This is a widely held misconception, routinely repeated. If we were slowing so rapidly that clocks required an extra second per year, we'd have ground to a halt billions of years ago. This wrong information manages to gain acceptance because Earth *is* in fact slowing, thanks to the Moon's tidal pull on the oceans, which acts as a giant brake on the world's beaches. It causes the Moon itself to move away from us at the rate of 1.3 inches a year and slows Earth—but only by one five-hundredth of a second per century, a rate too infinitesimal to require any annual meddling with clocks.

Others say that leap seconds are needed because Earth's rotation is irregular, with tiny glitches that make it go a bit faster sometimes and a bit slower at others. That's true enough, and this does help us understand why we need extra seconds in some years but not in others. Yet if this fully explained things, why have we only had to add seconds and never take any away?

The fault is not in the stars. According to NIST's Don Sullivan, the necessity originated over a thirty-three-month period in the late 1950s, when Earth's rotation was carefully monitored to find the exact length of the day. (This was needed in preparation for the new system, where clocks would henceforth be kept in sync with the turning of our planet.) Apparently Earth either spun abnormally slowly during that span or the measurements were slightly off, because the day and year's official length has generally failed to exactly match our planet's spin. The year is nearly a second longer than the system anticipated. A philosopher once said that for everything there is a good reason and the real reason. In this case, the real reason for insertion of leap seconds is to keep our two systems of time measurement—Earth's rotation versus 86,400 cesium seconds per day—in sync with each other. Leap seconds may be a somewhat jury-rigged solution, but they do the job of keeping the whirling of our clocks and our planet in Teutonic lockstep.

Time-Space Reversal

Perhaps the strangest time concept involves black holes—places where, several prominent theorists now believe, space and time switch roles. Inside a black hole's so-called event horizon (the border beyond which no free-falling object can escape the hole's gravity) is a zone where space becomes time and time becomes space, they say. This means a hapless astronaut would be forced to keep moving forward in space as surely as we here on Earth are compelled to march onward through time. Theorists have demonstrated that this would be a logical consequence of the extreme warpage of spacetime in the immediate vicinity of that frozen, collapsed, ultradense singularity. (For more on black holes, see chapter 31).

Some relativists, however, such as Professor Tarun Biswas of the State University of New York, argue that this notion makes no sense whatsoever. They point out that being impelled inexorably forward until one hits the singularity in the black hole's center would then amount to one's time coming to a halt. And what could that possibly mean?

Obviously, time's actual composition remains as elusive as that of sausages. Maybe that's why unwrapping its mysteries is such a timeless pursuit.

• Chapter 30 •

Light-Speed Magic

At high speeds or in places with strong gravity, funny things happen on the way to work. Time does indeed slow down—not good if you already have a boring job—and distance shrinks as well.

Time dilation has been amply exploited by science fiction, but the warping of distance is less widely appreciated. It means that if you could travel at 99 percent of the speed of light, the universe would suddenly become seven times smaller than it is now. A star about seven light-years distant (like Sirius the Dog Star, which dominates the winter sky) would magically float just a single light-year away, reachable by your next birthday. A football field shrinks into a lot barely large enough to park a trailer.

The effect becomes amazing as you zoom to just under the speed of light. Traveling 99.9999999 percent of light speed, you'd experience a dilation factor of 22,361. Your clock would tick off just one year while more than 223 centuries simultaneously elapse back on Earth. In addition, all distances in front of you would shrink by this margin. You could reach the heart of the Milky Way in a single year. With that much territory at your disposal, your social life could expand enormously. You could hold your next party near the black hole at the galaxy's center, and toss peanuts in just for laughs.

Of course, there's no escaping the fact that when you returned the party would *really* be over. During your two-year round-trip 45,000 years would have passed on Earth. New societal mores and technological advances would have made our world unrecognizable. You'd be seriously old-fashioned, and we're not talking about whether double-breasted suits would be in or out. You'd be lucky if they didn't throw you into a zoo. (The 25,000-light-year round-trip to the galactic core would actually require more than two years. Sci-fi odysseys often ignore the issue of comfort. The journey would be more enjoyable if the acceleration were gradual enough for you to survive the g-forces. Too much of a gung-ho attitude is a bad thing when an antimatter engine is under the hood. Push the throttle too hard, zoom to within a percentage of light speed in anything less than a week, and you'd be instantly converted to a smooth film of protoplasm on the back wall. To use real numbers, stretching out both acceleration and deceleration to a full year apiece would safely bring you to the desired speed with a steady one-g pressure, comfortably matching the everyday gravity your "You go, I'll stay here" buddies experience back home.)

Forget about sightseeing along the way. Another wrinkle is that near the speed of light everything in the universe will seem to lie directly ahead. When you're walking briskly through a rain shower, you must tilt the umbrella forward a bit to keep from getting wet. Chapter 2 noted that when you're driving through a snowstorm the flakes seem to come from ahead of you, while the rear window hardly gets hit at all. Light, too, is affected in this way: Because of Earth's motion around the Sun, the stars aren't quite where they seem to be, an effect called aberration. If we traveled faster, this effect would grow increasingly pronounced, until at just below light speed every-thing in the universe would shift until it all appeared dead ahead. In front of you, out the cockpit window, hovers a solitary blindingly brilliant "star," which is everything in the universe clumped into one dazzling ball. It is sur-rounded by blackness. Out the rear window you see—absolutely nothing. Because nothing is there.

Light and Information

Light uses a variety of methods to tease us with its outrageous changes. Its famous constant of 186,282.4 miles per second (or 299,792,458 km/sec) represents its speed in a vacuum, but, as noted, light moves more slowly in denser media, such as water (140,000 mps). It travels slowest of all through a diamond (which is why, in negotiating its facets and internal reflections, light's component colors bend in various directions to give that gemstone its coveted brilliance).

In 1905, Albert Einstein gave meaning to a wild observation made during the previous few decades by the Dutch physicist Hendrik Lorentz, the Irish physicist George Fitzgerald, and others. They had all realized that light travels at a constant speed and they also understood how profoundly remarkable this is. When a snowball is thrown at you, it hits you harder if you're running into it than if you're dashing away: common sense. But when particles of light emitted by your car's headlights strike a highway sign in the distance, they hit that sign at the same speed whether you're going a hundred miles per hour or you're parked. In a vacuum, light's unwavering speed of 186,282.4 miles per second is invariable, making it a true "constant."

Imagine if wind acted the same way. It would then feel like an unvarying gentle breeze regardless of whether you were stationary or holding your arm out the window of a fast-moving vehicle. So right out of the starting gate, light is peculiar and unique.

Of course there's also the issue of what, exactly, light is. In the Standard Model of How Things Work, light's fundamental unit is the photon, a force-carrying particle, a visible sign that electromagnetism is doing its thing. Photons are created when an electron jumps inward to an orbit closer to its atom's nucleus, releasing energy, and they vanish when they've met and excited some other electron. Encountering an object, a photon acts like a massless bullet that weighs nothing and whose energy depends on its color. Blue light is more powerful than green or red. And all visible light is more powerful than electromagnetic radiation at longer wavelengths, such as microwaves.

While photons are en route to their destinations, they can best be visualized as waves of energy, magnetic fields oscillating at right angles to electric fields. Each field creates the next and the combined electromagnetic wave zooms along at its famous speed.

Still here? In recent years, various researchers have made headlines ("FROZEN LIGHT!") by succeeding in slowing light down. Big deal! We've always known that slowing light was possible. Light automatically slows down when traversing any transparent medium. As noted during our morning commute in chapter 2, sunlight going through your window glass decelerates to about 120,000 mps and then instantaneously speeds up again once it's through. Particles can even exceed light's velocity in one or another medium. Electrons jazzed by powerful magnetic fields near some massive stars (the synchrotron process) can be hurled through a gaseous nebula faster than light can traverse it. When that happens, a beautiful blue shock wave called Cerenkov radiation is created as the particles break the "light barrier."

Under certain conditions, such as in material with extremely high optical densities like Bose-Einstein condensates, light can be brought to a near halt. The "frozen light" business is actually the process of "imprinting" a photon's characteristics (such as its wavelength) on the material— characteristics that are then released at a later time when properly stimulated. But the original photon does not actually hang motionless in space, like a cartoon character momentarily hovering beyond the edge of a cliff after realizing he has gone too far. Still, photons have been slowed to as little as thirty-eight miles per hour—not even fast enough to keep up with freeway traffic. Through it all, light's sovereignty in a vacuum has never been seriously challenged.

But here comes the shocker.

It starts with the odd world of quantum mechanics, a realm that makes Alice's adventures a tame stroll in the park. Quantum theory is a mind-twisting funhouse in which tiny particles can simultaneously exist and not exist, only to spring into reality the moment someone takes a look. Hold on: It gets worse.

A number of physicists accept the "many worlds" explanation for quantum phenomena. They hold that there is a separate universe for each option. The idea is that if you measure an electron you've forced it to make a choice as to what state it's in, where it exists, or how it's moving. Actually, that's not accurate: *It* hasn't made a choice; rather, *you* have suddenly joined the universe where it exists in the state you observe it to be in. Another "you" inhabits a separate universe, where you observe the electron in a different place or state. (So maybe you really did marry that cheerleader or football quarterback in an alternate universe.) Theorists unhappy with so many simultaneous realities have created equally far-out alternatives to the many-worlds explanation, but none have been widely accepted.

In 1935 a trio of top physicists—Albert Einstein, Boris Podolsky, and

Nathan Rosen—wrote a now-famous paper in which they addressed an aspect of quantum theory that was bizarre even by quantum standards. Examining the prediction that particles created together ("entangled") and then separated can somehow know what the other is doing, the eminent trio argued that such linked behavior must be due to "local" effects rather than some sort of "spooky action at a distance" (a quote cited often thereafter in seminars and lecture halls). The paper was so celebrated that synchronized quantum antics became known ever since as EPR correlations.

In that rare instance, Einstein was wrong. In 1981, three French researchers performed astounding demonstrations of these correlations that led to a new wave of reinterpreting quantum mechanics. The pace picked up in 1997, when a professor of physics at the University of Geneva created pairs of entangled photons and sent them flying apart along optical fibers. When one of them encountered the device's mirrors and was forced to make a random choice to go one way or the other, its entangled twin, seven miles away, always acted *in unison* and took the complementary option.

"In unison" is the key phrase here. The reaction of the twin was at least 10,000 times faster (the experiment's testing limit) than it would have taken light to traverse those seven miles. So how did the twin "know" what its twin was doing? The behavior of the pair was presumably simultaneous. Indeed, quantum theory predicts that an entangled particle "knows" what its twin is doing and instantly acts on that "knowledge" *even if the pair live in separate galaxies billions of light-years apart.*

Bizarre? Momentous? That, and more. The implications are so enormous that they have driven some physicists to a frantic search for loopholes. In 2001, NIST researcher David Wineland announced the elimination of one of the main criticisms of the French and Swiss experiments—the objection that they had failed to detect a sufficient number of particle events and that this had introduced a bias by causing observers to preferentially see only those twosomes that acted in unison. As reported in the journal *Nature* that spring, Wineland used beryllium ions and a very high detector efficiency to observe a large enough percentage of the in-sync events to seal the case. No arguments. This fantastic behavior is a fact. It's real. But how can one material object instantly dictate how another must act or exist when they are separated by large or even unimaginable distances? This one's a lockbox with no key. Few physicists think that some previously unimagined underlying interaction or force is responsible. Asked what he thought about this paradox, Wineland expressed an increasingly accepted conclusion: "There really *is* some sort of spooky action at a distance." Of course, he knew that this clarifies nothing.

So there it is. Particles and photons—matter and energy—apparently transmit knowledge across the entire universe. Instantly. Light speed is no longer the limit. Physicists say this does not violate relativity because we cannot send information faster than light, since the "sending" particle's information is governed by chance and not controllable: How can you send data, if the medium (whether light or sound) is "doing its own thing" and not under your influence? However, we're walking a fine line about what constitutes "information." *Something* is being conveyed instantaneously, and the implications will resonate throughout science (and philosophy) for decades to come.

How Fast?

Despite many heroic attempts, nobody could determine the speed of light until a few centuries ago. It was just too fast to measure. The Danish astronomer Olaus Roemer finally managed it in 1675, when he explained why the Jovian moons all bewilderingly sped up in their orbits whenever Earth was heading in Jupiter's direction. Obviously, said Roemer, the images of Jupiter's moons reach us sooner in that circumstance because their light doesn't have quite as far to travel as it did a moment before. The net effect is to make them appear to whiz around Jupiter at high speed, Charlie Chaplin fashion. Simple, really. And knowing the rough distance to Jupiter let him nail down a figure for the speed of light—well, within 25 percent. (Another nice touch of nature: Jupiter's three innermost large moons orbit in a synchronized 1-2-4 ratio; that is, Io makes four orbits as Europa makes two and Ganymede makes one. They move like a precise marching band, and this choreography is visible even through steadily braced binoculars.)

These days we pin down light speed with an apparatus like a rapidly rotating polygonal mirror that catches a light beam and reflects it to a measuring device. Light is fast, but you can comprehend it—its speed is far from infinite. That we can visualize the fastest thing in the universe is in itself remarkable. In fact, exploiting its velocity for our own purposes has become almost routine. Research facilities bounce laser flashes off reflectors left on the Moon by the *Apollo* astronauts; the delay between the pulse and its echo from the Moon is always about two and a half seconds; precise timings let us measure the Moon's changing distance from Earth to an accuracy of an inch! Light's 2.5-second Earth-Moon round-trip equals the half-million-mile distance that an average American drives in thirty years. So it's not beyond our grasp. Traveling at light speed for one second, you could go around Earth's equator nearly eight times, a truly nauseating carnival ride. Going that fast for a thousandth of a second, you could commute from New York to Washington. In a millionth of a second, you could go from the bottom to the top of the Empire State Building. And in a billionth of a second (a nanosecond), you'd travel 11.75 inches, essentially a foot. That's the useful statistic, because it means that everything around you is seen as it was in the past by as many nanoseconds as it is feet away. A pitcher ninety feet from the batter observes him "connect" ninety nanoseconds after it has actually happened, then watches the crisply hit ball increasingly move into the present moment as it approaches his glove. This makes the ball seem to move faster than it really does. (On paper, that is; it's of course not noticeable.)

Since light sets the ultimate speed limit, and no image or information can exceed that velocity, we can never know what anything is like *now* elsewhere in the universe. The delay grows ever greater as we look farther and farther, and it equals the age of the universe when we peer 13.7 billion light-years away. This, then, is the *edge of the universe,* beyond which nothing can be seen. (To be even more specific, energy started streaming freely through space 380,000 years after the Big Bang, and energy from that moment would be the oldest that could possibly be detected. Since stars are the primary source of visible light, and the first star formation began 200 million years after the Big Bang, that would be the oldest light we could hope to see with our eyes rather than perceive with microwave-detecting instruments. The most ancient possible observable star is thus 13.5 billion years old.)

There's no technological way of getting around this limit, although the EPR correlations do tantalize. We view stars in the Andromeda Galaxy as they were 2 million years ago and there is no possible way to get their current news. Nor, watching us through some supertelescope at this moment, could Andromedans see anything but Earth before the existence of *Homo sapiens.*

For the same reason, weapons that employ light will strike an enemy without warning. No technique, no amount of clever design could alert you that a pulse from an energy weapon was on its way until it actually arrived. Sci-fi movies show spaceships swerving to elude lasers or photon torpedos. Lots of luck: Such evasive maneuvers will always remain fictitious.

Then there's the redshift and blueshift business. While each photon of light hits you at precisely the same speed whether you're crashing into it head-on or racing away from it, your travel does succeed in making it change color. (In demonstrating spectroscopy to college students, and observing the three colorful emission lines given off by the glowing argon/mercury vapor in streetlights, I've often wished they could take turns running toward the light while looking through the spectroscope. It would be exciting and enlightening to see the yellow component turn green, the green turn blue, and the violet line vanish as it shifts into the ultraviolet. Alas, most of them seem too sluggish to run fast enough to experience this blueshift for themselves.)

But our high-velocity future astronaut will indeed watch the universe change. Sealed within the cocoon of her spacecraft, her face won't change when she looks in the mirror, but neither the position, appearance, time, length, nor color of the universe will match the view of those back home.

Travel at close to the speed of light will let future travelers go anywhere in the cosmos while hardly aging. And all this without violating any rules of physics. So it's not necessary to reach Warp 7, or exceed light speed, or do any of the no-nos prohibited by science. We can obey the law and still get there, even to other galaxies.

The only disadvantage attached to distant high-speed travel: Earth is eons older when you return. Your descendants have evolved. They're no longer human. Your jokes don't get a laugh. You want to tell everyone about Andromeda, but your words sound like monkey grunts to them. The new earthlings eventually just shrug and eat you.

A bad day by our present standards. Still, to witness eons of Earth's history, wouldn't you board the Light Shuttle just the same?

Chapter 31

Black Holes:
Twisted Space, Frozen Time

The violent scenario that produces a black hole is fearsome stuff, surefire drama, enough to generate countless magazine articles and science shows. The sequence: A heavy geriatric star collapses and as it shrinks its surface gravity quickly increases, compressing it ever smaller. The implosion deflates the unfortunate sun like a leaky balloon, until the velocity needed to escape its surface exceeds the speed of light. Then, *wham!*, it blinks out. If light can't escape, neither can anything else.

The runaway collapse continues until the star achieves infinite density and occupies no space at all. It has become a singularity—a dimensionless point—and is surrounded by an "event horizon," an invisible spherical shell just a few miles across, beyond which anything falling inward must vanish without hope of escape. (Despite bad press, black holes do not suck up stars or planets. When they feast at all it's on subatomic particles or, at most, an occasional intact atom.) If our own Sun collapsed to become a black hole, which is not possible, Earth would continue to orbit it just as before. We would not be pulled in; in fact we wouldn't experience the slightest increase in the Sun's "pull," because its mass would remain unchanged.)

If an incompetent astronaut crossed that event horizon, he wouldn't particularly know it. No physical barrier is there, no signposts; nothing really changes at that moment except that his phone calls are no longer returned. Then, quite abruptly, he is yanked violently downward and torn into goulash. An inglorious end to what until now had been an exciting flight.

It's a familiar script, presented often and in graphic detail. But here's a surprise: It probably never happened and never will.

The explanation for that (and for so much else) begins with Einstein, who proved that time and distance can twist and change. As noted in the preceding chapter, if you travel fast or hang out in strong gravitational fields, your time runs slower. Astronauts on the International Space Station, which is orbiting at 16,000 mph, have slower-running clocks than those on the ground. But *Apollo* astronauts on the Moon had the opposite happen; they grew old faster because they were in a weaker gravity than their stay-at-home friends. It didn't earn them early retirement benefits, but their speed-up was real enough to have been measurable had they brought superaccurate clocks along.

Muons are another case of time/distance distortion, except that these routinely kill people. Created when ultrafast cosmic rays collide with atoms of air some thirty-five miles up, muons live for less than two-millionths of a second, which even at their near-light velocity lets them travel only a half mile before they vanish. So they shouldn't live long enough to reach Earth's surface.

Yet they do. About 240 muons pass through your body each second and they're not totally harmless; occasionally one will strike the wrong bit of genetic material in a cell nucleus and cause one of the seemingly spontaneous cancers that have always haunted our species. Done in by a muon. Another bad day. Muons make it all the way to the ground because their time is slowed by their high speed, delaying their metamorphosis into harmless electrons. Or at least, that's our take on it. From a muon's perspective, things are very different. Time passes normally, and it is distance that is compressed. To a muon, Earth's atmosphere is just 1,000 yards thick, shallow enough to penetrate in those couple of microseconds.

Whose perspective is right and whose is wrong? That's Einstein's point: Both views are valid. Neither our sense of time and distance nor the muon's is absolute.

All of which is prelude to black hole theories and counter-opinions. From the perspective of the collapsing star's atoms, there's no problem about being yanked into a violent compression to form a singularity of infinite density. (Well, maybe there is. Some relativity experts, like SUNY's Professor Biswas, insist that relativity doesn't take into account factors like the strong nuclear force, which could become increasingly influential as material approached the singularity stage. They argue that since infinite density and zero volume are meaningless concepts anyway, it's not any *more* meaningless to suppose that some process like the strong force might halt a star's collapse and prevent a singularity from ever forming.)

Even if singularities can develop, from our perspective and the time frame of everything outside the black hole (meaning the entire universe), the collapsing material's time freezes so thoroughly that it takes infinitely long to crumple into a singularity and produce the accompanying event horizon. And an infinite amount of time has not yet elapsed, and never will.

So black holes cannot form, period. The only ones that might possibly exist would have had to be already present at the Big Bang, before time began. The common garden variety collapsing-massive-star scenario so often presented simply cannot happen in our reality, our universe, our time frame.

This is precisely the point I debated in 2001 with one of the country's top astronomers; let's call him Alan Phillips. (Edited out is everything I said that fails to make me seem a genius.)

> PHILLIPS: You're going to tell readers that black holes don't exist? You shouldn't do that! Just because we can't see them doesn't mean they're not there. Lots of things exist that can't be seen, like extrasolar planets.
>
> BERMAN: But those are real, and someday better instruments may see them. Black holes will never be detected, because it takes infinitely long for them to form and infinite time has not yet passed since the universe's birth.
>
> PHILLIPS: It's a matter of perspective. Muons see time and distance differently from us, yet their viewpoint is just as valid as ours. Similarly, a singularity forms in its own time frame even if it doesn't happen in ours.
>
> BERMAN: There's a difference. Muons exist from all perspectives. Only their paths through time or distance are seen differently by them and by us. But black holes never form at all in our time frame.
>
> PHILLIPS: But they *do* form.

BERMAN: Really? Then cite a possible date in our reality in which a supposed black hole like Cygnus X-1 might have arisen. Yesterday? A billion years ago? Day two after the Big Bang? When could such a thing have happened!

Science and human experience always describe things in Earth's time frame. No textbook would claim that our atmosphere is just 1,000 yards thick simply because a muon sees it that way. Why, then, are we supposed to negate our reality and that of the whole universe in favor of the perspective of an infalling atom in a collapsing star? At the very least, our time frame—in which singularities and event horizons never form—should be accorded equal, if not exclusive, validity, which is not the case today in science articles about black holes.

Continuing the debate:

PHILLIPS: OK, I'll agree that singularities and event horizons have never formed and never will. But people would be confused if you said that black holes don't exist. How about instead redefining a black hole? Call it a thin membrane of dense material that hovers just above an event horizon.

BERMAN: You mean, above where an event horizon would form in the imaginary infinite future? Fine!

And we shook hands.

Observationally, this changes nothing. We'll still observe the howls of flickering X rays from trapped infalling material. We'll still call the areas of immense compact gravity "black holes." The reappraisal is more philosophical or visual. Instead of picturing sinister vacuum cleaners, think of black holes as floating trash compactors—bizarre spherical membranes of dense material that hover in space as time grinds to a halt.

Density in Extremis: Cygnus X-1

Now that *what* we call a black hole is clarified, consider the *where*. The most famous specimen hangs over our heads once a day as Earth rotates beneath it. From much of the United States it's straight up or nearly so—a collapsed star named Cygnus X-1, source of a concentrated beam of flickering X rays that would dazzle any superhero who possessed X-ray vision. It's a place that represents one of the extremes in a universe awash in super-stuff.

Yawning emptiness makes up most of the cosmos. But here and there lie pockets of nothing's opposite. Thanks to a quirk in the laws of gravity, situations arise in which old massive stars crush themselves smaller and each bit of additional collapse creates a stronger gravity that, in turn, accelerates further collapse. Where does it stop?

Our Sun is not massive enough to end as a black hole. In its old age, some 5 billion years from now, it will stop compressing when it shrinks to the size of Earth. Then we'll have 300,000 Earths' worth of material packed into a sphere called a white dwarf, hundreds of thousands of times denser than iron. Each chunk the size of a sugar cube will weigh as much as an automobile (giving new meaning to the term "compact car"). The collapse stops there only because the electrons in each of the white dwarf's atoms can't be forced too near each other; they need their space and manage to resist further breakdown, a process called degeneracy pressure. White dwarf stars are all around us in space. Sirius, the Dog Star, has a white dwarf companion visible through good backyard telescopes—that is, when the space between the whirling duet appears large enough, as it will for several years toward the end of this decade.

If a star is at least 1.4 times as massive as the Sun (like Sirius itself, which is more than twice as heavy), its greater mass produces so much gravitational force that electron pressure can't halt the runaway collapse. The crunch doesn't stop until the shrinking sphere is just twelve miles wide. Imagine a ball weighing 500,000 Earths and small enough to fit between Exits 1 and 2 on the freeway. Yet such bad things happen even to good stars. The situation produces neutron stars (see chapter 21), whose material is so dense that a piece the size of a sugar cube weighs 100 million tons.

But all that density is spare change compared to the black hole that daily passes overhead. To achieve *that* kind of crush, Mount Everest would have to be packed into a ball no bigger than an atomic nucleus! Each black hole fragment the size of a Ping-Pong ball contains an entire planet Earth. That's the minimum packing job necessary to achieve enough density to warp space completely around like a taco. No wonder we have strange dreams with this thing floating directly above us each summer night.

Popular media depictions of black holes feature their supposed ability to suck things in permanently. Is there really no escape? That's an old question for physicists. Even leaving Earth without falling back down is no easy exercise. A cannonball fired upward always returns. But with a greater explosive charge and a higher speed, the missile goes farther. Long ago, physicists calculated the velocity that an object would need in order to leave Earth permanently: It was (as we saw in chapter 4) 6.9 miles per second, and this became known as our planet's escape velocity.

There's nothing complicated about either escape or velocity, yet together they often confound; even many science teachers don't fully comprehend the concept. Suppose someone had a rocket that could generate only wimpy amounts of power but could do so continuously, like the weak but steady ion engine that propelled NASA's *Deep Space 1*, the first spacecraft to use ion propulsion in flight, to a comet in 2001. Given Earth's escape velocity of seven miles per second, could such an anemic rocket ever wrench itself free from our planet's gravitational glue? Many people imagine that it could not, but even a speed of one mile an *hour* is enough to permanently leave your in-laws behind. That's because escape velocity applies only to a one-time, one-shot speed—a single push or thrust, like that cannonball—not to something enjoying the benefit of continuously applied force. If a rocket shuts off its engines just above our atmosphere and coasts from that point on (which is how we've always gone to the Moon), then it must indeed have a faster speed to get away, because it's steadily slowing. But if the engines keep firing, it could climb away at any lethargic speed it wished and still succeed.

Fast forward to black holes. Aside from the issue of whether or not they form in our reality, there is the question of whether or not whatever-it-is-that-is-there can be escaped. Could we someday drop in to see for ourselves?

Even some esteemed physicists feed the popular notion that there is no way to leave a black hole once you've crossed its event horizon. But again, that daunting greater-than-light-speed escape velocity applies only to unaccelerated things, like photons or bits of captured debris that don't have little rocket engines. Such natural objects really are stuck there forever. But there is no reason why a rocket with a reasonable sub-light-speed velocity couldn't break free and backtrack along the path it followed in. Geodesics (paths through spacetime) are always two-way streets.

The real problem for an adventurous crew is that the warping of space-time around a black hole ensures—from the outside universe's perspective—that an infinite amount of time will pass before the astronauts come back across the event horizon. If and when they do, from our perspective an infinity of time has simultaneously elapsed. The question becomes: What can the astronauts possibly see upon reentering our universe? To them, time has passed normally and their rocket has gone in and then perhaps speedily come back out. But with an eternity of time having meanwhile elapsed back home in our reality, what would the astronauts encounter? Here an old song comes to mind: "*Que sera, sera. . . .* The future's not ours to see." This might well be the real stopper for those exploring a black hole. Whatever experiences they had would forever remain their secret.

There's another potential problem with escaping: the popular hypothesis (see chapter 29) that space and time change their roles within a black hole's event horizon. Nobody knows whether this is true, but if it is, one *must* go forward to the singularity, just as one *must* keep traveling forward in time here on Earth by inexorably aging. But what happens to time at the singularity? We can imagine coming to a stop when traversing some distance in space, but what could it mean to come to a stop in time?

Light in the Dark

Researchers have recently found new ways to study these ultradense objects that play havoc with spacetime. In 2000, the Rossi X-ray Satellite observed odd oscillations—like a tuning fork pitched to a demented choir—generated by material orbiting a seven-solar-mass black hole with the catchy name of GRO J1655-40. Ten thousand light-years away, this "microquasar" has jets of atom fragments spewing like geysers at right angles to its accretion disk of infalling material. The sci-fi-like structure generates a 300-cycle-per-second X-ray emission—just what would be vented by orbiting debris a mere forty miles above the event horizon. An even more recent discovery of a second oscillation at 450 cycles indicates additional flotsam just thirty miles from the horizon. (Apropos of nothing, that oscillation cycle is just a bit lower than the whine of a mosquito, which at 600 cycles matches and is caused by the period of its flapping wings.) To orbit in semistable fashion that close to a black hole's event horizon requires that the entire spacetime continuum surrounding the black hole be in motion—which means that the black hole itself must be spinning. While every celestial body in the universe rotates, this is the first clear proof that black holes don't just make our heads spin, they do it to themselves.

In 2001, the Chandra X-ray telescope made the first discovery of variable X rays coming from a black hole (J1118+480) outside our own galaxy. Then the great orbiting instrument teamed up with three other NASA space observatories—Rossi, Hubble, and the Extreme Ultraviolet Explorer—to focus in unison on a strange black hole whose disk of accreting material halted mysteriously *far* from its event horizon. Instead of the twenty-five-mile limit that some scientists had expected (the distance at which stuff should have begun to vanish because gravity had redshifted its light beyond detection), a heating process may turn the doomed material into a hot gas bubble that halts the inward-spiraling material some twenty times farther away. Their observations have let us visualize something no humans are ever likely to see directly—the actual physical violence surrounding these ultimate cosmic mousetraps.

In any case, we are viewing either objects in the act of collapsing or else those somehow created before time began at the precise moment of the Big Bang—and not a nanosecond later. A fully collapsed black hole, one with an actual singularity, might well be the only kind of formation that (unlike stars, planets, and even atoms) predates the visible universe we see today.

Either way, the neighborhood of a black hole is an unforgiving place where the galaxy's clutter and riffraff are inexpensively removed. What more could we ask of it?

Chapter 32

Going to Extremes

On this cozy blue planet, many of us live in a comfort so habitual that we notice it only when it's interrupted. A power failure causes us to be temporarily cold, warm, or in the dark. But we rarely face lethal conditions—at least not the natural kind prevailing in the perilous kingdoms that rule the night's expanses. The brightest and largest objects and the most violent regions of spacetime lie light-years from our front lawns, where they neither impinge upon nor threaten our daily routines. But even at their unimaginable distances they serve to clarify the nature and limits of reality, the odd outreaches of the larger nursery from which we sprang.

Anyway, superlatives are always fascinating, the stuff of headline-grabbing news articles and gossip. We can always count on an audience for reports about the wealthiest, the smartest, the most perverse, the most powerful, the most egocentric individual: Average is much less entertaining.

Like definitions of contemporary morality, some of physical reality's edges are marginal or blurry. Nobody can define the diameter of the smallest or largest galaxies or the nature of the universe's two primary components, dark energy and dark matter. Other limits are razor sharp, not fuzzy at all.

In astrophysics, one of the best-defined boundaries is the coldest temperature in the cosmos, which is exactly 459.67° below zero Fahrenheit (-273° Celsius). Since what we call heat is simply atomic or molecular motion, and motion stops virtually dead in its tracks at this temperature, there simply cannot be anything chillier. In 1997, a distant cloud of expanding gas was found to absorb the meager trace of heat that permeates interstellar space, and in 2003 this gas cloud (the Boomerang Nebula, 5,000 light-years away in the constellation Centaurus) was confirmed by Hubble researchers as the coldest known place in the universe, even colder than the cosmic background and only a degree above absolute zero.

Well, perhaps not the coldest known. While most of the rest of the universe basks in the 5°F heat left over from the Big Bang (usually expressed in the Kelvin scale as 2.73°), the absolutely coldest place, so far as is known, is much closer to home. Not the Antarctic (where an impressive minus 129 was recorded on July 21, 1983) but in limit-pushing research laboratories that have achieved temperatures within three billionths of a degree of absolute zero. That was the only way to achieve the remarkable Bose-Einstein condensate, an entirely new state of matter. We've already seen that our knowledge of individual subatomic particles or even entire atoms is always limited; we cannot, for example, precisely know both an electron's momentum and its position; pin down one and the other almost magically becomes fuzzy, instantly. Well, what about chilling atoms to absolute zero? Wouldn't that do the trick? Since all motion stops at that temperature, we'd know the particle's momentum (zero) and we'd also know its position: right there, frozen in place. But Albert Einstein and Satyendra Nath Bose predicted in 1924 that quantum laws would prevail as matter approached absolute zero and nature would somehow create a state that would keep that information hidden. And indeed, in 1995, researchers at NIST and the University of Colorado succeeded in cooling 2,000 atoms to within twenty billionths of a degree of absolute zero. *Wham!* The atoms merged into a single blurry blob, a sort of superatom never before seen. We *still* couldn't pin down any individual electrons, because their individualities vanished in this new state of existence, the Bose-Einstein condensate.

Traveling the other way, upward, there are no limits whatsoever, since temperature has no theoretical ceiling. The Sun's surface is a mere 11,000°F (6,000° C), while blue stars, the universe's hottest suns, deliver surface heat of over 40,000°F—an energy that takes a million years to worm its way outward after escaping the stellar core's unbelievable nuclear-bomb temperatures of many millions of degrees.

Even that unfathomable heat is dwarfed by a supernova. Such a billion-degree furnace—incomprehensible to our minds but meaningful to the enterprising sizzle of the universe—depends on such conditions to forge otherwise impossible-to-create elements, including those that make life possible.

If we treated the entire universe as a patient and kept recording its temperature, we'd find that its present reading of 5°F is only half what it was 8 billion years ago. The cosmos is getting colder all the time. All the warmth cooked up by its 200 billion visible and estimated 1 trillion invisible galaxies (each of which pops off a supernova once per century, on average) fails to compensate for the dominant chilling mechanism of the universe's runaway expansion.

Which leads to the next superlative: speed. This is another easily defined margin, because nothing with any mass can quite reach the speed of electromagnetic radiation. Paradoxically, while the fastest objects are perfectly well defined, the slowest-moving are not. Velocity must be stated relative to some fixed point, some reference, and we have no stationary grid system against which motion can be compared. Thus a snail on Earth's equator still rotates around our planet's axis at 1,038 miles per hour, whizzes around the Sun at 66,000 miles per hour, heads toward the Andromeda Galaxy at 100 miles per second, and makes who knows how many other movements relative to various galaxy clusters.

In our own cosmic subdevelopment, the slowest spinning object is Venus, whose equator crawls along at just four miles per hour. A person could jog around that planet's waistline and keep the Sun from setting. As for motion of an object through space, the winner so far is the newest member of the solar system, Quaoar, a billion miles farther from the Sun than Pluto. At just under a half-mile per second, this tortoise's giant circular path carries it around the Sun only twice per millennium.

The Sun calls up another easy superlative: the most luminous object in human experience. It delivers 40 billion times more light to us than the Dog Star, Sirius, the night's brightest. The Sun's brilliance is intense enough to produce retinal lesions in anyone foolish enough to stare at it for more than a minute or so (not merely for a few seconds, as eclipse-o-phobic news media often claim). Like Lot's wife, we have been forbidden to let our eyes linger on the most riveting thing in the neighborhood.

But there are suns in the galaxy nearly a million times brighter than ours; the "most brilliant" prize must be awarded far beyond our solar system. Rigel, the well-known star in Orion's foot, shines with the light of 60,000

Suns. Not impressed? Then how about the spectacular and enigmatic S Doradus, visible from near the equator and all points southward; it has a luminosity of half a million Suns. Not yet bright enough? In 1997 the Hubble Space Telescope detected a star 25,000 light-years away, partially hidden behind clouds of dusty hydrogen near the center of our galaxy. Blazing with at least a million times the Sun's brilliance, it has been dubbed the Pistol Star, and it releases the same amount of energy in six seconds that our Sun emits in a full year. It is the all-time, all-star winner.

Maybe not. A supernova is the straight flush that beats a full house. It outdazzles any star. In 1987, the supernova in our companion galaxy the Large Magellanic Cloud radiated 500 million times the energy of the Sun! In the year 1006, Oriental records tell us that such a temporary object dominated the heavens, appearing some hundred times more brilliant than Venus. It cast distinct shadows on the countryside even from its perch 3,000 light-years away, emitting 100 billion times the light of the Sun. And that's number one in the temporary luminosity department.

As for the farthest object visible to the naked eye, that's the oval cloudy smudge in the constellation of Andromeda, visible from rural skies every moonless autumn night. Looking dim only because it's so distant, it lies far beyond the stars of the Milky Way. (Note for nitpickers: A slightly farther galaxy named M33, in the faint constellation of Triangulum not far from Andromeda, is marginally visible to keen eyesight from extraordinarily dark sites.)

By the way, Andromeda displays the same almost edgewise (13°) slant to us that we do to them. That Andromeda lies so close to the Milky Way in the autumn sky is proof that we must be nearly edge-on to any Andromedans looking our way. It's almost as if there were a mirror between us and each galaxy saw the other as though looking at themselves.

For "most energetic" body, the winner, excepting any healthy two-year-old, is a gamma ray burster. One of them pops off daily (on average) in distant realms of the cosmos. In a single second each unleashes as much power as will be generated by our Sun in its entire lifetime. Such short-lived explosions dwarf even supernovae and may be caused by collisions between neutron stars—exceedingly rare events in today's spacious universe but more plausible way back in the long-ago violent era from which their light has come to us.

In the quick-pulse-of-brilliance department, our final superlative award comes back to Earth. Our brightest artificial light is the 10-billionths-of-a-second, 1-trillion-watt laser, invented in 1995.

On to the longest and shortest intervals of time. The longest known period (other than in a doctor's waiting room) is the rotation cycles of galaxies. Since these are recurring, repetitious events, they are clocklike, but their ticktocks are barely perceptible. If our Milky Way's spin is representative, a typical "galactic year" is 240 million Earth years.

The opposite measure, the shortest possible time period, is theoretical but much more precisely delineated. It is the Planck-Wheeler time, named for physicists Max Planck and John Wheeler, who showed that any interval shorter than a ten-millionth of a trillionth of a trillionth of a trillionth of a second becomes meaningless, since it would then be impossible to know which of the events it divides came first and which afterward.

For rarest naturally occurring element, there's only one candidate: uranium. Surveying the cosmic inventory, for every trillion hydrogen atoms you'd stumble across 80 billion helium atoms, 740,000 oxygen atoms, and 450,000 carbon atoms. But just a single atom of uranium.

The smallest items must be found in the zoo of subatomic particles, such as electrons and quarks. But while still labeled "fundamental," it would surprise no one if quarks ultimately prove to be composed of yet smaller stuff. Obviously, largest objects are simpler to classify, because they're easier to see and far less likely to have their status usurped. In this category, red giant stars such as Betelgeuse or Mira take the prize. They're so enormous that on a scale model portraying each as a hot-air balloon twenty-five stories in diameter, Earth would be the period at the end of this sentence.

Of course, collections of stars are even larger, so that galaxies become the biggest discrete items known, and the bloated, inflated kings of galaxies are the giant ellipticals. The nearest of these is M87, an elliptical galaxy at the heart of the Virgo galaxy cluster 50 million light-years away. Light itself requires a million years to cross from one end of M87 to the other. This behemoth also boasts the largest black hole known, with a mass of 3 billion Suns.

The fastest spinning natural objects, on the macroscopic level, are pulsars. The famous Crab pulsar rotates every thirtieth of a second, but more than a hundred less celebrated pulsars spin hundreds of times per second, with 860 revolutions per second the current record holder. It's dizzying simply to imagine living on the surface of such an object. If one could survive its gravity—which would crush any visitor instantly, spreading her remains smoothly around the surface like a film of oil until no part stood more than an atom high—the spin rate would still be lethal. Imagine: The stars of the night sky would be luminous lines instead of points of light, whizzing from horizon to horizon in less than a thousandth of a second.

Except for the coldest temperatures (which, as noted, are created in laboratories), none of the other extremes of the universe are found anywhere near our planet. We have been born and nurtured in a Shangri-La, as far from the edges of violent physical reality as is possible. Someday, knowing what and where they are, later generations may venture to explore or even exploit the extremes of the cosmos.

Such over-the-top conditions may well be fathomed by our minds, but conceiving them is nothing like encountering them. They will surely remain as foreign as a symphony on Saturn or laughter echoing through the sea. Can anyone contemplating them, even safely from afar, doubt that ours is a strange universe?

Appendix:

Answer to puzzle, chapter 17:

The ages are nine, two, two.

Reasoning: The woman revealed to her friend both the product and the sum of the three daughter's ages, which should have been enough information to deduce their ages. (We, the reader, weren't given all that information, but the woman was.) Yet she says it's not enough. Why? That's the crux of the puzzle. Her claim of a *lack* of information supplies the critical information we need!

Looking over the possible combinations of three factors that yield 36, we find: 36, 1, 1 (which includes a set of one-year-old twins); 18, 2, 1; 9, 2, 2; 12, 3, 1; 6, 3, 2; 6, 6, 1; 4, 3, 3; and 9, 4, 1. But which of these is correct? The first woman ought to know, since unlike us, she was also given the sum of the ages. Armed with the sum, she could narrow it down to one set of ages. The fact that she now says, "That's still not enough information" can have only one possible meaning: Two or more of the above combinations must have the same sum! Looking them over, we see that indeed, 9, 2, 2 and 6, 6, 1 both add up to 13. So the first woman still isn't able to narrow it down from those two possibilities, until she's told that "the oldest one has blue eyes." Only 9, 2, 2 contains an oldest *one.*

Acknowledgments

Quirky revelations about distant realms and Earth itself have been central themes in my published articles for decades. While most of this book consists of entirely new explorations, I am also delighted to now revisit topics first presented in some of my monthly columns in *Astronomy* (1999–2003) and *Discover* (1990–2003) but which space restrictions prevented me from fully developing. I am grateful to various editors at *Astronomy*, and particularly Corey Powell of *Discover*, for their behind-the-scenes improvements and face-saving intercepts. In a few instances I've also elaborated on ideas first presented in my book *Cosmic Adventure* (1998, William Morrow).

For countless look-overs and key suggestions, I am especially grateful to Paula Dunn. For considerable help in whipping the manuscript into final form, thanks go to Sara Lippincott. If this book proves a smooth and pleasurable read, Dunn and Lippincott are the behind-the-scenes heroes. However, it would be a strange universe indeed if the present work's first edition is completely free of missteps despite their efforts. Here mea alone is culpa.

Finally, for having faith in this project and in me, my thanks go to John Sterling and David Sobel of Henry Holt. And to the dapper and wonderful Al Zuckerman.

Index

About the Author

Bob Berman is a contributing editor to *Astronomy* magazine and writes that magazine's "Strange Universe" column. He is also a monthly columnist for *Discover* magazine, and the astronomy editor of the *Old Farmers Almanac*. He is the director of Overlook Observatory in New York State and adjunct professor of astronomy at Marymount Manhattan College. His previous titles on astronomy include *Cosmic Adventure: A Renegade Astronomer's Guide to Our World and Beyond* and *Secrets of the Night Sky: The Most Amazing Things in the Universe You Can See with the Naked Eye*. He lives near Woodstock, New York.